IMHO (In My Humble Opinion)

.

IMHO

(In My Humble Opinion)

A guide to the benefits
and dangers of today's
communication tools.

RJ Lavallee

bent spoon
Multimedia

Boston / Walnut Creek

Cover by Stephen Sheffield

ISBN: 978-0-615-25988-8 (paperback)
ISBN: 978-0-615-26230-7 (hard cover)

To:
My Parents for their encouragement,
Atticus Fisher for keeping it regular, and
my wife, Heather: the best partner a man could ever dream to have.

For:
My boys, Tanner and Ethan.

Contents

Prelude

How did this happen?

When my oldest son turned five years old he desperately wanted a Nintendo DS portable video game. My wife and I were not big gamers, though we were both exposed to, and influenced by games like Pong, and the first Atari home systems. Video games came into being when I was a child – while I was still in elementary school – and they were mystical boxes of entertainment. I can remember so vividly sitting in the basement of my friend's house playing Pong: the wheels controlling the video paddles; the sound of the ricochets; the simple graphics; its straight white lines floating on a black background, which at the time were so novel.

But the mysticism surrounding these boxes, and cords, and wheeled paddles, that made rudimentary two dimensional icons move up and down on the screen, intercepting and "hitting" a moving square of light, punctuated by tinny computer generated music, was not enough to keep me from Lincoln Logs, Legos, Whiffle Ball games, and flashlight tag. As I grew up so did the video games. In my teen years games like Tetrus and Asteroids captivated many kids, just not me. That a few of my friends, a few of my close friends, were spending more time playing games like these spurred me to bug my parents for an Atari system of my own. The Atari 2600, after all, defined the threshold of the technology phenomena that would soon engulf our society. Priced at $199, it sold one million units in 1979. Coupled with the popular arcade game, Space Invaders, two million units sold in 1980. (Wikipedia, 2008)

Christmas came in 1981 at my house, and so did the game, but no one in my family would play with me. The friends of mine who had gotten Space Invaders at their homes on their Atari systems the year before had returned to playing Whiffle Ball and Cream the Carrier, instead of wanting to come to my house to play the games they'd already tired of.

Three weeks after getting the Atari, we returned it and exchanged it for a camera that I put to very good use over the following years. My parents were not disappointed by my decision to trade in the Atari. In my hazy recall of those hormone filled early teen years, I seem to remember that they moved more quickly than I had expected when I said I thought it best to take the gaming system back to the store. If I remember correctly my mother was already standing at the back door to the house with the Atari neatly packed in the box, asking me "so you'd rather get that camera?"

Years later the decision to have a camera as a pre-teen would benefit me in many ways and hinder me in others. I carted my new Pentax K-1000 everywhere. I learned how to develop my own photographs in the school darkroom, and spent hours winding my own canisters of film, looking through the lens, snapping pictures, developing photos, and generally feeding my teenage narcissism, fueling my immediate self-importance, yet creating a record that I could later look back upon and laugh.

Having a sister four years older than me also had an influence over where my focus was. She was the cool older sister, already in high school, getting ready for college, who had the pulse on avant-garde art and music. The influence of older siblings is important. My friends who also had older siblings were the ones drawn away from the video game craze, while my friends who were the oldest siblings were more inclined to stay engaged with video games, either at an arcade, or at home. A miniature digital divide was already building as I became frustrated with those friends who would rather play video games, and started spending more time with the friends who were not so drawn to this new technology.

Would keeping that video game console, however, have

changed my life that much, made me that much of a different person, or was my decision to exchange the console for a camera the changing factor in my life? Of course the importance of the moment of exchange of one for the other is hyperbole; it was, however, a memorable moment. The importance, I recognize in hindsight, was the birth of the digital divide. My older sister, just under five years my senior, couldn't be bothered by video games. Actually, she was very bothered by the sounds, and flashing lights, and couldn't see my brief attraction to them.

My parents, who were already accustomed to their children liking things they didn't understand, humored my interest, but made it clear they would not go so far as playing the games with me. I was intrigued by the games, but not enough to sit and play them on my own – something a real gamer would have no problem doing. To be honest with myself I wanted the Atari system mostly because my best friend had one too. Ironically, however, his little brother played the games more than he did, and we never played the game when I went over to his house; we played Wiffle Ball outside with all the other neighborhood kids.

Another subtle but memorable moment in seeing the growth of the digital divide happened a couple of years after college when I was sitting in the apartment of a couple of friends who graduated a few years after me. Amazing the difference of a few years. These guys were glued to Madden Football. I watched them playing this game and was...intimidated.

After trading in the Atari, I rarely played video games. I'd like to say it's because I didn't want to spend the money, since except for an Atari at home access to video games was at arcades. By the time I was in high school, arcade games (video and at-home) had become ubiquitous. No longer were arcades the venue of pinging and dinging pinball or claw-retrieval games, the electronic whoop, whoop, whoop of video games echoed from every arcade hall.

I didn't play because when I played I was never that good. The sense of urgency in most of the games shot my adrenaline through the roof, and I was no good at controlling it. The same lack of control I had when playing second base in a baseball game – fielding a ground ball with a runner going from first to second, having my mind scream "YOU'RE GOING TO GET THE GUY OUT," which invariably lead to me bobbling the ball – this same inability to control the adrenaline rush was unbearable as video games became more advanced and intense. At least in the baseball game the play would be over quickly, I could breathe, and regain composure, taking each little experience to build on itself and slowly train myself to focus on the moment, and finally, by high school, make the play without the bobble.

In the video games, I could never figure out how to slow myself down and focus; I didn't have the patience. If I ever had to go to battle for my life I'd either be a Tasmanian Devil, or be the first shot, or both. The moving targets in video games. The increasing tempo in the music. I couldn't deal, and I'd go down in flames after a few short minutes – if it lasted that long – leading me to say "I'm no good at this," and leave it at that. But just as I had recognized through subtle cultural cues, and peer pressure, video games very quickly became part of the cultural lexicon of young boys and men. Staying just ahead of the wave of the growing digital divide didn't mean that I would be able to escape its influence.

Being born when I was meant I was able to pick and choose the side of the digital divide I wanted to live within. Siding with my older sister, as many of my peers sided with their older siblings, I shied away from video games, but was still intrigued by technology. I was one of the first in my class to use a computer. I convinced our middle school science teacher to purchase a couple of Tandy TRS-80s for our school, where a few of us taught ourselves how to program silly little computer applications in a language called BASIC. Ironically, we were programming games, and yet I had no interest in playing the games.

In speaking with my parents about this project, my mother reminded me of the times I would bring the TRS-80 home from school on the weekends. She didn't get it. She saw me tinker with this box that needed an audio cassette to save whatever it was that I did on it, and the black screen with the white blinking rectangular cursor, and she didn't understand what all the fuss was about. She saw the little Hello World programs I would write, over which I was so proud, using a random number generator to make the words "Hello World" appear on different parts of the screen, and she wondered why I was wasting my time with this oversized game. But games would become very important. Little did I know that my own familiarization (or lack thereof) with being immersed in video games would become a parenting issue thirty years after returning the Atari for the K-1000.

Before acquiescing to our oldest son regarding his Nintendo DS, I wanted to see what all the fuss was about around it. Did our son simply want this game system because all of his friends had one? Was he going to get the DS at the same time that his friends were getting bored with theirs? Were the kids who had these DSs the types of kids I wanted my son hanging around? Were these games leading kids to spend more time camped in front of a game, anywhere, anytime, sitting on their butts, not getting outside to do what was more familiar to me and my wife: running around with balls, and riding bikes?

These questions were what lead to the production of ***IMHO***. My wife and I were talking about story ideas, and she asked, "What about looking at how video games are influencing kids today?"

At first all of my questions revolved around video games. The more I pursued the questions about video games the more my questions pointed to Technology. Yes, with a capital "T."

After all, just like culture, Technology is always changing. Communication is changing too, but human's desire to communicate has not. Very early on in my research of this project I discovered one woman, danah boyd [sic], with whom I wanted to speak. She's a recent recipient of her PhD from the iShool at UC Berkeley. She's a giant in the field of examining how "teenagers socialize in networked publics like MySpace." She's in such demand for speeches, and interviews that one of the few places I was able to connect with her was in Milwaukee, WI before she spoke at a conference.

Granted, the trip was more self-serving than flying to just see her, as I was also able to coordinate an interview with an author in Chicago who I quote here later, as well as wrangle a weekend visit with an old college friend. With all of this technology, however, the irony is one of my more coveted and revered interviews occurred with terrible lighting and a lot of background noise in a coffee shop in downtown Milwaukee. Such is the lot of a documentary filmmaker working with no budget.

During our interview danah quoted Allen Kay, saying, "Technology is that which is invented after you were born."

This statement sets up one of the foundations of cultural generation gaps. After all, children brought up when a specific technology is introduced are involuntarily immersed in that technology.

A term danah defined for me, and one I heard in previous and subsequent interviews was "Digital Natives:" the kids who know how to use a computer better than their parents, whose thumbs are glued to a cell phone, who expect things from technology, are impatient with technology, and who are rarely impressed by new technological breakthroughs. Throughout *IMHO* we will see different authorities present different perspectives on what new Technology is bringing to our culture. We will also learn how the development of Technology does not happen in a vacuum: culture and Technology influence each other.

IMHO is not a story documenting how technology is developing; it is about how technology is facilitating communication. The process surrounding producing *IMHO* is about what technology is allowing content creators. With only a single plane ride to the Midwest, and numerous drives from the East Bay of Northern California to Palo Alto, and San Francisco, *IMHO* was put together...virtually. Nothing is committed to film and video tape. The camera was a new high-end consumer high-definition camcorder that records straight onto a hard drive. *IMHO* only exists in my head, and in my computers. But here you are consuming my virtual product.

Thirty years ago Alvin Toffler in *Future Shock* noted "A strange new society is apparently erupting in our midst." (p. 10)

This comment and the book attempted to address the growing cultural divide that many blamed, and still blame on Technology. From *Frontline* to *USA Today* Technology is continually blamed for social ills and the unraveling of our culture. Welcome to the future, where the people screaming that Technology is going to be our downfall may just be right.

IMHO doesn't look at all of Technology; it can't. It sticks to a large umbrella of technology called Computer Mediated Communication (CMC), which in today's Western society means everything from the cell phone to the laptop. One element of CMC that I will visit is a place called the Metaverse, a term coined by Neal Stephenson in his science fiction novel *Snow Crash*. The Metaverse in the book was a virtual world where people lived and worked – a place that was just as significant and real as their terrestrial lives. I will not argue here whether science fiction inspires technology, or technology inspires science fiction. Stephenson's book is an important reference point, however, because it has been the inspiration for the development of much of the modern Metaverse, just as so many other elements of science fiction from *Aliens* to *Star Trek* have influenced the directions technologists have attempted to move.

Other elements like cell phones and blogging may seem unrelated to fantastical 3D virtual worlds, but they are much related. After all, many digital natives will update their Facebook pages, or blogs from their cell phones. The cell phone has become the ultimate tool for communication, facilitating phone-to-phone communication, phone-to-website interactivity, and email communication. No longer does a digital native need to be tethered to a desktop or laptop computer in order to be connected. Using very little imagination, you can begin to imagine the huge amount of possibilities this un-tethered means of communication may open. The possibilities are both frightening, and thrilling.

Jamais Cascio is another resource whom I feel privileged to have met. As Jamais says in his bio on his website, he "writes about the intersection of emerging technologies, environmental dilemmas, and cultural transformation, specializing in the design and creation of plausible scenarios for the future." Jamais is a futurist. He takes an academic and humanistic approach to seeing where life is taking us. One project in which he was asked to participate was called the Metaverse Roadmap Project, sponsored by the Acceleration Studies Foundation. Collaborating with industry leaders, academics, and other futurists Jamais helped pen the *Metaverse Roadmap Project Overview*, a 28 page synopsis illustrating to where the 3D Web may evolve. The 3D Web, the Metaverse, is possibly the most compelling, frightening and exciting aspect of new technology cresting our societal horizon.

At a recent conference where Jamais Cascio was discussing the deeper nuances of the *Metaverse Roadmap Project Overview*, he supplied a scenario where you are walking down the street wearing a pair of digital sunglasses – wirelessly connected to the database of your favorite social networking site – that intuitively draws lines of connection between you and others: degrees of separation, as in the play *Six Degrees of Separation*. Now imagine that behind the scenes the database is also linked to a facial recognition database. Now imagine the sunglasses are also digitally scanning your environment, providing you a digital display floating in your line of sight, providing you information regarding your surroundings. Now imagine you've instructed your glasses to paint over the people in your environment with whom you do not want to interact...for any reason. We quickly go from plausibly mind blowing to micro-fascist in one small step.

Perspective is everything when looking at a future moving as quickly as ours. To examine different perspectives, **IMHO** interviewed Virtual World Industry professionals, academicians, and children (the voice of the future) to draw conclusions that are bound to make you at least scratch your head, and at most become terribly intrigued when trying to draw the line between what is virtual and what is real.

Introduction

Money makes the world go 'round

"China's fastest-rising currency isn't the yuan [sic]. It's the QQ coin – online play money created by marketers to sell such things as virtual flowers for instant-message buddies, cellphone ringtones and magical swords for online games." – **The Wall Street Journal**, 2007 (Fowler & Qin, 2007)

Legend of Mir 3 is an MMORPG: Massively Multiplayer Online Role-Playing Game. These games follow the definition; they are massive – involving thousands of players simultaneously playing the games – and they are only playable through an Internet connection. Created in South Korea, Legend of Mir 3 is a fantasy battle game with compelling animations, and a back-story complete with stories of good and evil, and vanquishers and the vanquished. Players become a part of a guild, where they individually attempt to gain skills, and strengths, and work towards rankings that are published for all to see on the game's official website – which is not the game, just an introduction to it – and collectively work towards game-based goals. (http://www.lom3europe.com/) The *Guinness Book of World Records* certified that at one point 750,000 players were simultaneously playing the game.

In January of 2005 *BusinessWire.com* announced World of Warcraft – a similar fantasy-based game created in California – had broken records for the amount of users who were simultaneously using any online video game: 200,000 players were playing the game at the same time.

As recently as April 14, 2008, *GamesIndustry.biz* reported 1 million players were simultaneously playing World of Warcraft the month prior to the report. (Lee, 2008)

These numbers are massive. How can it be that 1 Million people in one country are playing one single game? There is a little bit of technical hocus-pocus going on behind the scenes. Computers have limits. Have you ever had too many applications open on your computer? It runs more slowly, and sometimes stops working. Computers run websites, and the same situation applies – if too many people are trying to use a computer that is responsible for displaying a website on the Internet that computer can potentially stop working. Computer scientists, therefore come up with all sorts of technical magic with the tools they have at their disposal. Terms like load balancing, and complicated algorithms attempt to manage how many people are using any single computer within the web of computers that make the website a reality on the Internet. These load balancing techniques mean that up to 1000 players can and will be in any one region of the game at any given time playing a single game, and when I say region, that refers to what the users of the game are seeing; they see themselves as playing within a physical space within the game – a physical space that doesn't physically exist. Or does it?

As with many MMORPGs (MMOGs, or MMOs for short), players pursue game-specific objectives, and within the game's scenarios the game presents specific items that hold value within the game. What does this mean? If you become the owner of a vest with magic properties, and there are 10,000 of those vests in the game, then you're holding your own in the game. If you become the owner of a rare mystical sword, the only one of which exists in the game, then you're not only a bad-ass, but the sword carries serious value within the game.

At the initial inception of MMOs that value stayed "in-world," leaving the value solely as something for barter, but something that still was very important to the gamers invested in these worlds. Interestingly, however, is that another well-known website was growing up at the same time that these MMOs were growing in sophistication: *eBay*. And because gamers are a curious group, people who thrill in the intellectual challenge of getting around road-blocks, a few gamers figured out very quickly that this new bartering platform called eBay could be neatly coordinated with a few of the MMOs to make in-world bartering lucrative in real-world currency.

> *Real money, itself, is already as virtual as the*
> *stuff that circulates in these worlds.*

This is a quote from Julian Dibbell, a freelance writer who has followed technology for years, specifically online worlds, has written for *Wired Magazine*, and a few years back published a book called *Play Money* that chronicled his attempts to make a living from doing nothing but trading virtual goods. He attempted to replicate the anecdotal evidence that people were generating enough real-world currency to not have any vocation other than buying and selling virtual goods. Those goods, in turn, had enough value to people who were willing to pay "real" money that the holders of those virtual goods were able to transfer the virtual rights to the virtual goods in exchange for real-world money.

This technical and mostly administrative leap from virtual wealth to "real" money also brings into question the veracity of the reality of our currency. Even before we talk about Internet-based video games, we have to ask ourselves, "What is real?" After all, the concept of paper currency, and even more of electronic transfer and our current monetary system is, ultimately, virtual. But to keep our focus on the MMOs, there was an MMO that found its users taking advantage of the eBay connection early on; it was called Ultima Online (UO). This was the same MMO in which Julian made his stab at trading virtual goods. UO was one of the first visually-oriented MMOs, and it followed the classic fantasy board game formula: a pseudo medieval environment, with related challenges, quests, and goals.

This MMO not only had users turning virtual items into real world currency, but it sparked a cottage industry, one that many might find ethically dubious. In areas of China, and a few other developing nations, savvy gamers started "gold farming" operations where the owners would find skilled gamers to play UO for ten to twelve hours at a time in exchange for room, board, and menial wages. Within UO the participants could perform simple tasks that would earn in-world "gold." The operator of the gold farm would then take the earned gold and offer it for sale on eBay, turning the virtual gold into real world money.

UO had, and still has external markets where UO currency (Brittania Gold) is traded just like any "real-world" exchange market. And just like the real world, fortunes are made and lost based on the fluctuations of that currency market: virtual currency arbitrage. As of August 2008, there is even one online firm, Playspan, which facilitates the bartering of virtual goods between unrelated virtual worlds. When Jullian Dibbel was in the midst of his **Play Money** experiment of making a real world living from trading virtual goods, the fluctuations of Brittania Gold exchange rates had a direct effect on the amount of real world money Jullian had at the end of a week to do real world things like pay the rent and buy food. Even with less consequential activities in virtual worlds like SecondLife.com that don't have the same distinct purpose as UO, Legend of Mir 3, or World of Warcraft, there is a market for Second Life's currency: Linden dollars.

Second Life is a highly chronicled virtual world, but not technically an MMO. There is no "purpose" in Second Life; it's a place where people can choose to either replicate their current lives online, or fabricate an existence for an alter-ego, each of which are pastimes with their own unique cultural interpersonal currencies. Corporate giants like Coke have a presence in Second Life, and CNN has a news 'bureau" in

Second Life. In Second Life you create a virtual representation of yourself, or a fanciful elf, or ugly troll. This digital stand-in is called an "avatar." What you do, how you look, the accoutrements with which you adorn yourself are up to your imagination, ingenuity, and the amount of Linden dollars for which you can work within Second Life, or that you can simply purchase at whatever is the going exchange rate, which is published daily on the website of Linden Labs, the creators of Second Life.

At the time of the growth of gold farming many MMOs forbade (and still forbid) the exchange of "in-world" objects for real world currency. Other MMOs like UO did not openly promote the exchange of goods for real world currency, but they did not prevent it. Human greed and capitalist ingenuity are crafty innovators. For the MMOs that forbid barter for real world currency, many also forbid players from creating in-world, automated characters that do nothing but accomplish tasks that create in-world goods of value. Remember, these are massive computer games, so savvy technical players can manipulate the computer code in the games to do things the inventors of the games never intended.

Chinese entrepreneurs recognized the low cost of their labor could create large volumes of in-world wealth without having to bend the in-world rules of automated processing of goods. At the end of the day, however, the reason this was possible was not because of places like eBay, but because of people: the gamers who would rather pay real world money for in-world gold, or other items of high in-world worth rather than invest the time required for earning and / or creating these items. The mechanics were the pure illustration of supply and demand: a free-market evangelist's dream.

But what does this really mean? For people who have grown up around technology making the conceptual leap is simple. This is an "oh, someone gave you money for that" revelation that seems fairly simple for someone who has always seen pixels on a TV or computer screen as synonymous with tangible goods. But for people, who understand tangible as being more tactile than conceptual, for those who have witnessed the birth of this technology during their adult lives, let's break this down a little further.

Imagine having a photograph. Whether you took the photo yourself, or someone else took the photo is irrelevant. Now, most people understand that you can take that photo and somehow translate it so that it lives inside your computer. The mechanics are mystical to many, electrical engineering and computer science theory to others, but from any perspective after some process or other you are able to now see the photo via your computer. Now imagine that you are able to allow tens of thousands of other people to see this photo, but not possess this photo. Welcome to the birth Digital Rights Management (DRM).

DRM is a simple concept with complicated consequences; it is a technical tool that restricts how easily a person can copy digitized art (music, images, movies, etc). It protects the people who create art, and the people who distribute it from piracy. What frightens businesses that sell creative content the most about the Internet is this difficulty in controlling what happens to something once it is digitized. DRM is not perfect.

Adding DRM is no guarantee that a savvy technologist won't find a way to still copy the content. Without DRM, once digitized, creative content producers have practically no ability to prevent anyone else from making an almost limitless number of copies of their creative content. In other words, policing the use of that digitized photograph – the one you put on your computer, the one that was a once-in-a-lifetime photo, from which you want to get credit for having taken, and for which you want monetary compensation whenever anyone uses it – becomes unwieldy in the age of digitization. Add to the equation the view that growing economic powers in the East have towards copyrights and ownership – their perspective on copyright laws is different from ours – and the value of digital information becomes clouded.

So imagine that you can offer this photo for sale to others: someone sends you money somehow (in the mail, through Electronic Funds Transfer, or some other means), and you then send the photo through the Internet in a way that the person can now posses the photo on their computer. So, you have exchanged a copy of this photo to someone for some form of currency, but now they have an easily reproducible copy, and you still have the original. Music publishers, movie studios, and book publishers all understand this process far too well, and given that their entire existence is based on making money from irreproducible copies – or at least from copies for which the process of reproducing is prohibitive to the general consumer – these businesses are looking to the future wondering how they are going to stay relevant.

Advancing the concept, in MMOs the photo in question is now an animated picture, which we will now call an "in-world item." What makes these in-world items irreproducible, which greatly increases their value, is how and where the items are produced; these are not made for transport or use outside of the virtual worlds in which they were created. You could say they need the virtual world they are in to exist at all.

Now we are only talking about other Internet users who have access to, and meet within one specific virtual world. Within this hypothetical world there are many, many in-world items. The users in the virtual world constantly fight for, play for, barter, create, and trade these in-world items. Within the virtual world these items have a lot of value, but the value is limited to life within the virtual world.

Let's return to the article from *The Wall Street Journal* referenced in the beginning of this chapter. Discussion over the relevance of virtual currencies is not over-due; it's practically after the fact. Virtual currencies are alive and well, and very significant. The Wall Street Journal article details how $900 million worth of virtual goods – pixels on computer screens – were traded in 2006. Just under half of those goods were traded in a single virtual world: Tencent, the virtual currency of which is the QQ. The company is traded on the Hong Kong stock exchange, is part of the fabric of popular culture in China, and attracts so much attention that companies like Coca-Cola regularly use the world of its own promotions. (Fowler & Qin, 2007)

This stuff is not Monopoly money, but it's nothing new either. What's the difference between trading virtual goods created or earned within a virtual world for "real" money and going to an arcade, pumping coins into the claw game, and working to take home a stuffed animal or other toy from your arcade exploits? That toy has value, can be traded for other toys, or even sold for real money, all from the spoils of an arcade game.

In researching for *IMHO*, my sifting through Google search results eventually brought me to Coye Cheshire, a Sociologist and Assistant Professor at the UC-Berkeley School of Information (iSchool). As he states in the bio on his school-hosted web page, his "work focuses on how various forms of exchange are produced and maintained on the Internet." After a couple of conversations over the phone and over coffee, I was able to spend an hour with him in front of my camera.

Regarding virtual goods he indicated that

> *...digital Information goods are very different in many ways than tangible kinds of things...A digital good might be a song, it might be something we create, a video... that we create for our own enjoyment. It could be a copy of a book, or an essay, and so if you...get philosophical about it, then you realize that there a lot things that we think of as physical goods, like a book, that when we read it our experience is really the way we thought about it and how we processed that – or a good song – something that reminds us of something else, that brings up a memory.*

In other words, when digital goods are eliciting the same memories / sensations as tangible goods, then the intrinsic value of those goods is no different than the tangible. If you think about music you have to agree that we are already there.

Julian Dibbell explains the bridge across the divide between simple digital goods and the importance of games – in whatever context you want to define games – but limits them to a digital environment. When the digital good is something that users make, the end result has many levels of pleasure associated with it including,

> *...look what I've done, and it's look what I've done better than that guy, better than any of these other people...It's competitive, and it's fun! That's the other thing that doesn't get talked about. There is a pleasure in a kind of self-motivated [process] that doesn't get talked about except for by, oh, Linus Torvalds [the creator of the Linux operating system, an alternative to Microsoft Windows] who actually points out that...this is the motivation, the [process] itself, is a kind of entertainment, a Ludic activity. So I would argue that...this phenomenon that I call Ludo-capitalism, this*

*broadening infestation of the global economy by play,
by Ludic energy, and motivation is bigger than we
already... might think it is, because it's already seeping
out and transforming things.*

Play as work is already here, and it's changing how people think about work and play.

Some people can easily walk away from games, even tactile board games like Monopoly and Dungeons and Dragons. Other people become very emotionally invested in games. For those very emotionally invested in the virtual world, these items take on varying levels of value. Opportunistic people put unrelated technological ideas together to create a new marketplace: the real-world purchase of virtual goods. While most virtual world transactions involve some process of meeting at a specific place within the virtual world, and then some process to exchange in-world items, savvy users figured out that an extra step in the process could be adding real-world currency to the process. Two virtual world participants would meet at their pre-arranged meeting place to exchange an in-world item, but the holder of that item would not give it to the other user until s/he saw a deposit in a PayPal account, or other real-world place to exchange real-world currency. So holders of valuable in-world items offer in-world items for sale on eBay, let the auctions run their course, gather the in-world information from the winner of the auction, collect real-world money for the sale of the in-world item, then meet the winner within the virtual world to exchange the in-world item.

Currency takes on new meanings. Currency is no longer Euros, and Dollars, and Yuan, and Yen. Currency is not necessarily something you can fold and put in your pocket or wallet, though as Julian Dibbell noted the "reality" of currency has been in question for some time. The speed and ease with which currency can now be created, and what can transpire because of it – what can be shared, and bartered, and bought, and sold – opens doors of possibility never before seen in our modern society, if ever. The sheer scale at which all of this can and does occur – involving tens of thousands of people instantaneously – and the broad reaching effects from the costs of a teenager doing too much text messaging to a person losing their life's savings on a virtual real estate gamble are astonishing. But how dangerous is all of this really?

Section 1: Danger!

Chapter 1: Death Online

"'Game theft' led to fatal attack." ***BBC News*** March 31, 2005.

Lock up your children, the future is pouring through the Internet and it's dangerous. Cyber-bullies. Pedophiles. Stalkers. Thieves. The Internet is providing a new conduit for every conceivable ne'r-do-well. Don't be fooled. That you sit safely behind a keyboard does not mean you are safe.

When I began investigating online dangers as more than a passing curiosity I wondered how bad relationships on the Internet could be, and thought, "how dangerous are games, really?" Where was the first place I turned to research this? The Internet, of course. One of the first Google searches I did was "online game murder." The first result was not a titillating story of murders arising from the events of online games, but of a link to online games that involved murder as a subject, though I should not have been that surprised given the popularity of television shows like the *CSI* or *Law and Order* franchises. Scrolling down through the results were links to what has become a famous case online: the 2005 account of events that transpired within Legend of Mir 3.

The operators of these 3D worlds developed each of them with an intended purpose. In World of Warcraft, to more effectively accomplish in-world, game-driven goals, users form guilds in which there is a leader, and as you might imagine the same personality conflicts that arise in real life eventually arise in World of Warcraft guilds. Within the gaming community, running a successful guild is so revered that many gamers will place such an accomplishment on a resume. Further complicating the conflicts are the augmented personalities that come with each user. Within a 3D world like World of Warcraft each user chooses and customizes an avatar from a list of game-specific options. Orcs. Dwarfs. Humans.

As danah points out "You are never going to be an orc...People often do choose races and roles that are reflective of certain aspects of their affiliations and identities."

And as recent research points out, if someone spends a lot of time, some may say too much time, in-world, the behavior that their avatar exhibits in-world begins to bleed into their real-world. The user starts to forget which world is their real world. The most obvious example of this comes from studies that Professor Jeremy Bailenson is conducting at the Virtual Human Interaction Lab at Stanford University. He was interviewed on NPR's *Weekend All Things*

Considered on April 27, 2008 about where he described the relationship between our demeanor and various visual cues in our everyday lives. There is a little bit of a proverbial chicken-and-the-egg question inherent in many of his studies. For example does wearing a well-tailored power-suit make you feel more confident in a meeting, or did you feel confident and therefore don the suit? What his studies have shown is that people who were supplied attractive avatars in a virtual world experience viewed themselves as more attractive in real-world interactions as much as an hour later. (Bailenson, 2008)

A month later *Time Magazine* published an article further expanding on Bailenson's own research. The author of the article, Kristina Dell, that Bailenson's research illustrates how visual we are as creatures, and how easily visual imagery manipulates our perception of the world. She notes that his research has proven that as little as 90 seconds of interaction with avatars in a virtual world can affect changes in your behavior in the real world. We carry our innate personal traits into our virtual worlds. That does not seem like a surprise. What is a surprise, however, is that if we construct our avatar in a virtual world to take on an alter ego – a personality we may want to express but restrain ourselves from exhibiting in real life – then as we express that alter ego in a virtual world, after doing so we carry those alter-ego traits with us back out into the real world, and this does not require being immersed in a virtual world for a long period of time. (Dell, 2008)

The blurring lines between an in-world persona and real world persona are easier to lose after being immersed in a place like Second Life. Unlike World of Warcraft there is no fantastical game-centric goal within Second Life. The persona you create is at your own hands; it is not guided by an external game creator. Create and bring your avatar into Second Life and in many places you will find other avatars simply hanging out. This is one of the greatest criticisms of Second Life as there is no purpose, but on a very metaphysical level, how much different is that Second Life from our First Lives? We create our purpose as we go along through our lives; our purpose is not bestowed upon us at birth. Heck, isn't that the basis of so many religions? Why are we here? So the attraction to Second Life can be obvious if you find someone who is dissatisfied with the roles s/he has fallen into through the course of his or her First Life.

This isn't a tangent to support that these virtual worlds are a possible Utopic outlet for dissatisfied individuals; rather, this is an illustration that real life idiosyncrasies follow users into their virtual lives, and, maybe more importantly, their virtual idiosyncrasies follow them from their virtual lives out into the "real" world.

Greed. Hubris. Anger. These motivational elements do not get left at the threshold of virtual worlds. Relating this to the research on behavioral patterns, one can easily see the anonymous mask of an avatar can quickly amplify these traits in a user. People inclined to skirt the rules in real life, technically savvy people who also enjoy the gamesmanship of skirting the rules, can easily become more inclined to try even harder to bend gaming rules in their favor.

But just as young fiction writers quickly learn that it's often difficult to dream up anything stranger than real life, the simple free-market economics growing around MMOs could never account for the human element.

Returning to the 2005 Legend of Mir 3 event, Qiu Chengwei, a Shanghai gamer, had built enough experience in the game that he was able to acquire a coveted "dragon sabre [sic]." Mr. Chengwei had befriended another player, Zhu Caoyuan, who he let "borrow" the dragon sabre. The value of the sabre (both in-world and real world) was not lost on Mr. Caoyuan, who quickly turned around and sold the "borrow[ed]" sabre for 7,200 yuan ($967 US (2008)). (BBC, 2005)

With a cursory view of the events to this point, few in the U.S. would really raise an eyebrow to what Mr. Caoyuan did, except for the amount of money this man was able to fetch for something that is nothing more than 0s and 1s stored in a database on the game's website server somewhere in the ether of the Internet. What is not told in the events of the story is the human element of this transaction. In MMOs like Legend of Mir 3, UO, and World of Warcraft, players are more than mere participants seeking to further their own cause. Real relationships are forged within these MMOs. Trust. Camaraderie. Friendships. Phenomena that many of us relegate solely to palpable "real world" experience play out in these virtual worlds. And the more sophisticated these worlds become in their features, and the visual landscapes they present, the more "real" the experiences become.

All of this human interaction occurs despite the terrible job that games perform in creating a platform for human networking. Christian Renaud, the former Chief Architect of Networked Environments for

Cisco Systems, and now the CEO of Technology Intelligence Group, a consulting company specializing in analyzing and reporting on emerging technologies said,

> *There's this deep human need to connect...Here...you're paying money to go in a play a game that has a narrative and has goals and objectives, but yet people will get on there and form social clubs and just do that – so they use it as a how, and not a what. They'll get in and they'll have little coffee cloches, and weddings and funerals and so forth using that.*

While MMOs like World of Warcraft were begun as nothing more than sophisticated online games, entire communities and cultures have grown up around them. Relationships are born. Deaths are mourned. Jealousy. Envy. Pride. Joy. Real human emotions are truly woven into these games. Teenagers have spent so much time in these games that they failed to graduate high school. Adults have spent so much time in the games at night that they have lost their spouses.

The first academician I interviewed, in fact the first on-camera interview I conducted, was of Byron Reeves, a Professor in the Department of Communications at Stanford University. I was fortunate enough to see Professor Reeves speak at the first annual Virtual Goods Summit held at Stanford University in June of 2007, a conference held with academicians and virtual world gaming industry professionals examining the current and future benefits of and challenges facing virtual currencies.

As detailed in his bio on his school-hosted web pages, "Professor Reeves has published widely on such topics as children and television, physiological responses to media, attention, memory, and emotion, the history of media effects research, political advertising, television news, and multi-player interactive games."

In our interview Byron spoke to the realism of this new virtual existence.

> *The invention of media [games, online worlds, etc], the ability to create these virtual people and places is really a visual illusion; it's almost a trick on the brain. We can think our way to a conclusion that they're*

not real, but as far as the basic processing people are bringing to these events the default evaluation is, quickly, that they are real. They deserve consideration as if they were real people and places, and only after some thought can they be falsified, and we participate, really, in a willing suspension of disbelief.

One anecdote Byron tells illustrates how studies have shown that when players in an MMO see the avatar of another player in the MMO, and that other player's avatar turns and recognizes (waves, or in other ways greets) the first player's avatar, the player, the actual person experiences the same endorphin rush as if s/he had a friend across the street recognize and wave to him or her.

In answering my question about how we physiologically react when we see an avatar wave to us he answered,

The same neural circuitry is firing when you get a picture of somebody that's pretty close to reality as is firing when that actual person is present, so the brain chemistry, the automatic responses are indeed very similar. You may be able to think your way around that similar response if you spend a lot of time, if you are in a virtual world, someone comes up to you and says something bad, or does something you don't like, you may be able to think 'well, gee this is just media, these are just pixels on a screen. It might be controlled by a computer rather than another person. I shouldn't take offense. I shouldn't be excited.' But that's really a very difficult response – probably not anything that's by any means the default.

The emotional rush from recognition even occurs in something as simple as text-based blogs, when a blogger checks in on his or her blog and sees that people (friends or complete strangers) have read, or even better yet, commented on the blog.

People like to be recognized. Coye Cheshire personalized his view of this,

Part of the reason why I'm even a sociologist is that I find it fascinating how social we are as creatures; that is, we really depend on other people. We really rely on, we care about, we interact with, given the choice, we tend to hang out together in groups...On the Internet this still tends to play out...A lot of [social cues] are limited in the online sphere, but that doesn't make it any less important to us to be able to interpret one another, to be able to glean some information – limited it may be – but glean some information, infer something about the other people who might be contributing to [the same online communities we are].

It's a base, primal need that all humans have.

So when Mr. Chengwei discovered Mr. Caoyuan had sold the dragon sabre, this enraged Mr. Chengwei. You have to imagine that Mr. Caoyuan knew that selling the sword would not make Mr. Chengwei happy. As an afterthought Mr. Caoyuan offered to share the proceeds with Mr. Chengwei. What Mr. Caoyuan probably did not expect was that Mr. Chengwei would track down Mr. Caoyuan and kill him. *The China Daily* said Mr. Chengwei stabbed Mr. Caoyuan with "great force" in the left chest, which caused his death. (BBC, 2005)

Sadly, the story of Mr. Caoyuan's death is not the only example of events in a virtual world leading to real world, physical consequences. A June 2001 article in *Time Magazine* describes the real-life gangland beating of an online gamer. In January of 2007 the *Associated Press* reported on a 22-year-old who was caught in the web of an Internet-based love triangle. In January 2008 the PBS series *Frontline* expanded coverage of the case of cyber-bullying that lead to the suicide of a teenager in New Jersey.

Like much of the Asia-Pacific rim, South Korea is worlds beyond the United States not in only in the adoption of technology, but in the types of technology available to practically everyone. Even in 2001, as *Time Magazine* reported, there was a dearth of internet cafés in South Korea. In 2007 67.1% of the Korean population regularly used the Internet according to the Ministry of Information and Communication, Republic of Korea. This includes all of rural South Korea. (Levander, 2001)

Anecdotal evidence indicates over 85% of urban South Koreans had broadband Internet connections as early as 2005, where as of July 2008 "55% of adult Americans" had broadband access. (Horrigan, 2008)

South Korea has distinctive cultural elements that allow for the rapid penetration of certain games into the culture.

The *Time* article stated that there is a popular online game the design of which plays much into a gamer's psyche as it does into the nuance of Korean culture: winning and working in groups. The game is a place where peers can convene, but one that also gives players a mask under which they can express repressed desire. (Levander, 2001)

While this wildly popular game in South Korea was adopted first by the mainstream populace, in typical American cultural spaces, as danah boyd presents, the first adopters of technology are usually the self-proclaimed "freaks and geeks."

danah continues,

> *If you look at social technologies at every turn, the first adopters are always the self-defined geeks, freaks, and queers – the marginalized people of society – who desperately want to find others like them, and are seeking to have the sort of freeing of their physical constraints.*

In the *Time* article, Paek Jung Yul is a gamer in the popular game described above, which is called Lineage. So did Mr. Yul know when he did what he did in the game that it would lead to the real-world consequences he experienced? The South Korean society involved in the game Mr. Yul was playing knew, or had knowledge of the real-world identities of most of its participants. Mr. Yul killed a character within the game, and he knew what he was doing. He not only knew what he was doing, but he boasted about it. What he was boasting about was that he had killed the online character of a locally know gang member. (Levander, 2001)

Eventually the gang member and four of his friends visited the Cyber café where Mr. Yul played, pulled him into the restroom, and gave him a real world beating. (Levander, 2001)

The first reaction might be "How terrible that events in a game, and a virtual one at that, would lead to real world violence." Then again, reading into the story a little more, you might ask if Mr. Yul was asking for trouble. But, again, he must have known what he was doing since the virtual killing, and the real world beating are now part of the lore of Mr. Yul and his friends, a badge of honor of sorts. (Levander, 2001)

Virtual worlds, real people, and real deaths are not limited to the realm of fantasy worlds where elves kill ogres. Nothing is more human than love and sex, and these are the two elements of humanness that pervade virtual worlds. From Match.com to Second Life, social networking sites to hardcore pornography, early adopters of the Internet quickly saw the possibilities this new conduit provided for connecting with other people, and the potential benefits of connecting anonymously.

What today's digital natives understand is that even though hacking is possible, most Internet sites provide enough reasonable roadblocks that allow you to maintain anonymity on the Web. We understand that most typical users of social networking sites use the Internet to actively broadcast who they are, posting pictures, and enough other evidence for others to quickly discern real world identities, but if a user uses a pseudonym, does not post a picture, and is cognizant of limiting the posting of other personal information, tracking down a user's real world persona can be easily hampered, but it is not impossible.

As adults quickly realize, life is difficult enough to manage when you are telling the truth. When a couple of not so innocent mis-truths are involved, imagine the potential struggles. A 2007 *Associated Press* article illustrated how quickly anonymity can unravel. Starting with the line "they were two middle-aged people carrying on an Internet fantasy based on seemingly harmless lies," you can begin to imagine what was going on in the online world of these two people that eventually lead to the death of a 22-year-old man.

After all, despite the display of pictures and seemingly personal information on social networking sites or other virtual spaces, as the article later quoted the Erie County Sheriff, who warned that when you are talking with someone on the Internet who have no idea who that person on the other side really is.

Without much effort I found examples of how an Internet

relationship can go bad. A 47 year-old, former Marine, and married father of two in upstate New York portrayed himself as a young Marine bound for Iraq. He was using this persona to correspond with whom he thought was an 18 year-old woman, but who was actually a 40-something mother living in West Virginia. Coincidentally – tragically – a 22 year-old who worked at the same place as the 47 year-old former Marine began chatting online with the same woman.

The online relationships progressed. The woman sent packages to the former Marine's home. His wife intercepted one of the packages, which contained lingerie and a picture of the woman's daughter. The former Marine's wife replied to whom she thought was an 18 year-old woman, ending the fantasy for the mother in West Virginia, ruining the former Marine's online relationship, and his real life marriage.

Unfortunately for the 22 year-old man, he continued the online relationship with the mother in West Virginia, unaware that the woman was really a 40-something mother, and unaware that she had been having an online relationship with the 47 year-old former Marine and coworker of the young man. The young man started boasting about the relationship he was having with the woman, who was also still chatting online with the former Marine. The former Marine was easily able to realize that his young coworker was having this online relationship with his former online lover. This enraged the former Marine so much that he eventually shot the 22-year-old in the neck as the young man climbed into his truck after getting out of work. (Thompson, 2007)

Love was complicated enough before the Internet.

One of the most tragic illustrations of the new Internet-based, social networking dangers facing digital natives specifically is the story of a 13 year-old boy who, after months of cyber bullying he committed suicide on October 7, 2003. The story received a lot of media attention, and was featured on a January episode of *Frontline* on **PBS**.

Kids are mean. Even if you are not a parent, most of us can remember the days on the playground, or even worse, the awkward days in middle school when fitting in, or at least finding a clique that suited your personality was paramount. The worst case scenario was for the kids who did not neatly fit into expected, prescribed molds. Jock. Geek. Nobody. What if you were part of a clique that made it easy for you to be a target of teenage vitriol? The words are painful. The sentiment is hurtful. And yet, the process has gone on for years. At least before the days of the Internet a child could escape from taunting and torment by simply going home.

Human beings are resilient, particularly when they know an end is in sight. School only lasts so long, and many children get through middle school and high school knowing that eventually they'll be able to go to college, or be done with school, at which point they can either reinvent themselves, or seek respite somewhere else. As the 13 year-old entered middle school, cell phone text messaging and social networking sites like MySpace were ubiquitous for early teenagers and pre-teens.

This boy was affable – not reclusive – and of course subscribed to the virtual communication media of his peers, understanding the importance of their social capital. Little did he know that participation in these media would also create an inability to ever escape the taunting and teasing from bullies at school. The cruelty of children runs deep, and can be arbitrary. The bullies targeting the 13 year-old found some reason to taunt him by calling him names like fag, and worse. Their virtual attacks were relentless, and eventually the 13 year-old relented by taking his own life. (PBS, 2008)

The tragedy in the story is in how the 13 year-old's parents handled the events leading up to the suicide; they did what practically any other caring family would have done. They were involved. They knew as much about their son as any parents of teenagers do, and yet they were unable to protect him. Of course there may have been behavioral cues that the 13 year-old's parents may have been able to react to, but seeing such cues in hindsight is easy. After all, most of us with children are not trained mental health professionals. We come to the parenting job supplied with the cultural cues of our own youth, not of today's youth – hence the perpetual generation gap – but maybe now, more than ever, parents need to be that much more understanding of the landscape in which our children are growing.

The influence the virtual world is having on people's lives is already very, very real. From purported gaming addiction to actual deaths, the line between the lives people lead online and the ones they have in "real" life are already blurred

Chapter 2: Lost in the Internet

"Experts Say [online] Gaming Can Be A Compulsion As Strong As Gambling." *CBS News* July 3, 2006.

"The downside is that Web 2.0 may be destroying civilization." *The Sunday Times,* April 22, 2007.

The word "Avatar" holds deep significance in Hindu and Buddhist lore. The first definition of avatar in the American Heritage Dictionary is "the incarnation of a Hindu deity, especially Vishnu, in human or animal form." (Dictionary, 2006) Wikipedia succinctly defines avatar as the following: "In Hindu philosophy, an avatar (also spelled as avatara), most commonly refers to the incarnation (bodily manifestation) of a divine being (deva), or the Supreme Being (God) onto planet Earth." (Wikipedia, Avatar, 2008) This is a very dramatic word for something that is a digital representation of either one's self, or one's plaything in an online world, and an equally compelling word for something that becomes the icon for a real person's immersion in a virtual world.

The danger of immersion in a virtual existence is not limited to immediate physical harm. But before we get to how harmful it can be, what is immersion? In Stephenson's **Snow Crash** immersion involved "goggling in" to a virtual world – a 3D, computer generated world in which user's avatars traveled, interacted, and conducted business. "Goggling in" refers to wearing a pair of wrap-around glasses, or some form of eyewear that limits your field of vision to that which is presented by the game. This concept has been important to virtual reality simulators since before the days of the movie **Tron**: the 1982 science fiction film where the protagonist is literally a part of a video game. Like virtual worlds today, in **Snow Crash's** virtual world there was no death, there was no decay. Once someone created something in the virtual world it was there until the managers of the virtual world decided to remove it, or until the computer it was created on ceased to operate. There were rules governing an avatar, that may allow an avatar to "die," but once dead it had the ability to return alive to the world.

This science fiction rendition of immersion in technology paralleled the time during which the novel was written. VRML. (Virtual Reality Modeling (or Markup) Language, pronounced VIR-ml) refers to an Internet programming language first discussed in 1994 that many thought ten years ago would quickly take over the Web and thrust us into a ubiquitous land of virtual worlds. Virtual reality goggles. A human's interface with technology was very limited by the technology, but imagination is not limited. Within the virtual world of **Snow Crash**, the human interface with the technology – goggling-in – may seem cumbersome, and not so advanced since we have that today, but one specific technological breakthrough in the novel is still ahead of its time: facial expressions. Facial expressions will make an avatar able to express more of those subtle visual social cues that make the nuances of human interaction quintessentially human.

Immersion in today's terms involves more of a suspension of disbelief than it does wrapping the field of vision in goggles, or forcing a computer to attempt to recreate intricate facial expressions. The work of people like Byron Reeves illustrates that the human brain is wired to relate to relatively basic visual cues. Immersion in a virtual world, therefore, does not require elaborate efforts; our brains are already wired to immerse ourselves in visual experiences.

On July 3, 2006 **CBS News** reported through a report associated with **WebMD** that kids were becoming obsessive about gaming. (WebMD, 2006) This statement came from Keith Bakker, director of Smith & Jones Addiction Consultants, who has created programs in response to what he sees as a growing problem in young men and boys. After all, our stereotypes of hard core video gamers usually hold true. The primary participants are young men and boys. Being the father of two young boys, this was one of the notions that first drove me to pursue producing **IMHO**.

In the report Bakker explained that the escapism provided by video games was not very different from the escapism sought and found by cocaine addicts. (WebMD, 2006)

For those of you afraid that video games will consume your young boys, leading them to become high-school drop-outs unable to constructively contribute to society, just wait, they're not the only ones about to fall into gaming's seductive grip. As far back as 2004 **USA Today** reported results from a survey that showed how the average age of video gamers was increasing. The average age of players was 29 years old, and the average age of buyers was 36, with 59% of the participants being men. This statistic, however, was limited to console games – the kind of stand-alone video game systems that plug into your TV set at home. What this statistic does not show is how online games have been increasingly attracting younger players, as well as more women. (Reuters, 2004)

Away from the easy-to-spot online games like World of Warcraft, similar games like EverQuest, or beyond media targets like the console game Grand Theft Auto lay casual games, and games that don't even appear to be games. The research company **Research and Markets** reported in 2004 that half of the visitors to gaming sites that were part of well-known web portals like Yahoo and MSN, were adult females. And these were not insignificant gaming sites, they were attracting as many players as World or Warcraft was in 2005. This is a big deal; it's akin to a drug dealer figuring out how to start selling his junk to the one demographic who has been trying to keep his steady customer base off of drugs.

As recently as August 2008, the increasing popularity of video games has taken off with girls as well. Gaming designers have finally figured out how to make games that appeal to their previously untargetable audience. With 174 million gamers in the US, this is a big deal. (Ivan, 2008)

There are cell-phone-based games. And the makers of the games and gaming systems understand how the online / interactive games are changing the marketplace – bringing non-stereotypical gamers in as potential customers. Anecdotally this element of the gaming industry was starting to wane until the introduction of the iPhone 3G.

A recent **CNBC.com** posting detailed how the introduction of one product – the new iPhone – will most likely resuscitate a lagging mobile game industry segment. This also signals how much farther away from the classic gaming console (think back to the Atari 2600) that gaming is moving. (Pisani, 2008) Who would have imagined playing computer games without being tethered to a wall jack for a power supply, and that playing those games would eventually allow for the interaction with multiple, if not multiple thousands of people without being tethered to a telephone line?

Sony Interactive is working on a 3D online world called Sony Home that will initially be offered and accessed through Sony's PlayStation 3 gaming systems. The importance of this move by Sony is not just their recognition of the importance of online / interactive games – games that allow users to play with multiple people at the same time – but with making the gaming console no longer a stand-alone box that only plugs into the home TV. That line has been blurring for a while now. For the past fifty years television broadcasters were the controllers of multi-media communication coming into the home. This communication, like the radio, was a one-way path. Apart from the radio, the telephone was the only communication medium that allowed for long distance, two-way communication between two people not in the same room. While cable companies first challenged and changed the way television broadcasters did business, the greater influence of cable companies came with the introduction of broadband Internet access.

Without going into the arduous technical details, you can think of information like water. More water, more information. Telephone lines allowed for a garden hose of information to enter and exit homes. Fiber optics and other inventions increased the telephone's capability to that of a small fire hose. Broadband communications from cable companies placed the water main straight into the home. While not a limitless flow of information, the opportunities this greater flow of information provided programmers were enormous. This invention, however, also meant there was more than one option for bringing and taking information to and from your home. Free market capitalists love it: competition.

This competition is leading to a further blending of all of these communication media that were once all very distinct and separate. Television stations stream their broadcasts on their websites. Websites stream archived television shows, movies, and radio broadcasts. Telephone service providers now offer Internet access. Offerings like Sony Home will leverage these technologies and continue to blur the line between what is online and what is not.

Anyone with children understands the marketing juggernaut presented by terrestrial products like NeoPets and Webkinz. NeoPets (part of Viacom) and Webkinz (Ganz) are both tangible, purchasable products, as well as virtual world brands. A child can either go on to the NeoPets or Webkinz virtual worlds to interact with their friends, or purchase the real-world stuffed animals these companies offer for sale. Each company has a different take on going about the business of this, but in either case they're driving their consumers – children – to participate with time and money in both the real world and virtual world spaces.

Insidiously, beyond the allure of the physical nature of these stuffed animals, are the online worlds these products drive children to. On the surface, the virtual worlds for NeoPets, Webkinz, and other child-oriented sites like Club Penguin appear harmless. Watch a child interact on these worlds and you see something far more troubling. Patterns develop with these children that drive them to want to collect more and more and more...of everything. More clothes. More furniture. More houses. More friends. Unknown penguins approach other penguins on Club Penguin and request friendship. Kids click "yes" faster than playing a game of whack-a-mole. I saw my son do this the other day. "Do you know who that is?" I asked.

"Nope," he disinterestedly replied.

What mattered to him was that his buddy list was growing faster than the coins in his bank, which was also important. More coins meant more things he could buy to put in his Penguin's igloo, or carry in his inventory list of clothes and accessories.

"But that is just virtual currency," you might add.

Yes it is just virtual currency, but in order for children to redeem their virtual currency from the Club Penguin catalog of items kids can buy with their coins – everything from different colors for their penguins to interior decor for their penguin's igloo – the children must be paying members of the virtual world. They can earn as many coins as they wish. They can play every game. They can participate in every aspect of this virtual world, except having full access to the catalog. They can see every page of the catalog, but unless the children arrange to have a paid membership the goods they can purchase with those coins remains tantalizing out of reach. Who would have thought that play money would ever take on such real meaning? When we were young you could use the disparaging term Monopoly money for something that was worthless. Play money is no longer that.

The children in these worlds are safe from predators – physical or sexual – but they are not safe from financial predators. The worlds are brilliantly crafted to encourage children to express themselves through their online characters.

In December 2007, fleeting rumors existed that Sony was going to bring back its Aibo robotic dog. The original Aibo was a small toy that almost looked like a robotic Snoopy. It could recognize up to 100 voice commands, and travel significant distances. And though Sony discontinued production of the robotic pet in 2006 amidst a corporate reorganization, the 150,000 who purchased the $2000 robotic pet were crushed that they could potentially not be able to find parts for their aging Aibos. Maybe Sony was premature in both launching and discontinuing the Aibo. The rumors about the Aibo indicate that not only would Aibo come back, but it would interface with Sony Home through both the PlayStation 3 consoles, and PSP. The Aibo would have its own avatar in the Sony Home online world, an avatar that would react to all of the real world experience the terrestrial robot-dog experiences...and the avatar could, theoretically, also influence the actions of the terrestrial Aibo; it would bring the virtual to the real...or maybe it's bringing the real to the virtual.

The concept is already in place with other virtual worlds. Back in Club Penguin children can buy a virtual pet called a Puffle; they don't have to be paying members for this. The puffle is a little animated puff-ball pet for the child's penguin avatar. Just like some of the first virtual pets – small hand-held toys with black and white LCD screens that its owners had to periodically monitor and push buttons in order to keep the pet "alive – the puffle requires food, and rest, and attention. Without all three of these, if the child takes his or her penguin avatar throughout Club Penguin while neglecting his or her puffle, the puffle will up and run away. Then the child needs to go back and earn more coins to buy a new puffle. Kids take this responsibility very seriously. My son was almost in tears when he had not been on Club Penguin for five days. He thought that the puffle would run away if he did not go online, not realizing that the programming routine behind the puffle only cared about the time that he was logged into the virtual world.

Webkinz takes the concept to the tangible – the half-step between the Club Penguin puffle and the virtually tethered Aibo. When a child buys this new stuffed animal toy – a Webkinz – the ear tag of the toy has all of the information a parent would expect to see on a stuffed animal's product tag...except for the registration ID. On each Webkinz toy is a unique identifier that already has the toy registered with a database tied to the Webkinz online virtual world – a world that is restricted to kids who have Webkinz. Well, anyone can register, but you will not have an avatar to travel through the Webkinz world unless you have a Webkinz pet with a unique identifier to register. Once registered, however, the Webkinz that the child has sitting on his or her lap takes on its own life – which the child controls – within the Webkinz virtual world. More brilliant still – from the perspective of marketing executives – is that to maintain access to already registered Webkinz a child must obtain and register a new Webkinz toy every year. Gold for Webkinz; nightmare for parents.

If we go back and take a quick look at Club Penguin, we see that it's a 2D world. There are no fancy 3D gaming graphics to draw in a child. Webkinz, similarly, is what industry people refer to as 2 1/2 D – there is the sense of perspective in its animation, like the basics from an introductory drawing class – but still there are no hyper-fancy graphics trying to make anything look realistic. Despite the animation styles these worlds use, children become easily immersed in these worlds all because of their suspension of disbelief.

We've seen that when a child purchases a Webkinz, the tag on the toy encourages the child to go online and participate in that virtual world. When a group of friends at an elementary school all have Webkinz, the toy encourages them to meet in this virtual world, to control their avatars, to live a different incarnation of either their selves, or what they perceive as the personalities of their Webkinz, and to congregate with the avatars of their friends. Think about Paek Jung Yul in South Korea and his friends in Legends of Mir 3. Groups of people – this is not limited to children – have the same sense of community regardless of the geography. Think of what Coye Cheshire said, "given the choice we tend to hang out together in groups." When one child has five close friends who all have Webkinz, and who all get together in the Webkinz virtual world after school, what do you think that child wants to buy, and wants to do?

The worlds are building a whole new generation of consumers. Brilliant and wonderful for a capitalist society such as our own, the value of currency – in this case online currency which the game makers have so far been able to prevent from being bartered on eBay – is obvious and dubious at the same time. In order to earn more in-world currency to purchase new products children must only complete gaming tasks. This pattern strains parenting skills, raising questions in young children regarding why earning money in the real world is so difficult. Within these worlds children are already involved in commerce. They may not be earning and spending US currency, but they are earning and consuming. Of course this is not new in the realm of video games, but this aspect of casual games has come closer to realism far faster than any computer graphics have in rendering realistic images.

While all the kids have to do is play small games like riding an inner tube behind a water ski boat while avoiding obstacles in the water to collect coins to purchase from the Club Penguin catalog, the kids quickly find which games they can play to maximize the number of coins they are earning in the time they are spending in-world. Some kids play the games because they like the games. Some kids play the games just because they are earning coins. At a very young age you start seeing the same base motivations that drive adults in their vocational choices. Is this good life training, or is this growing up too quickly, or is this setting unrealistic expectations for children that earning money is simply a game?

The influence of virtual worlds, however, is not limited to being online. The Nintendo DS is a great illustration of how children can become immersed in a virtual world that is not tied to the Internet. The handheld gaming system is designed and marketed to children. A unique feature of the product helps to expand the virtual experience. In games like Lego Star Wars, where the children control Lego-constructed avatars that look like the characters from *Star Wars*, children can also do a DS download, or "hook up." When they do this, the children must have a physical proximity to each other. An Infrared signal allows the children to play the game together, so each player sees the avatar of the other player in his / her game; they are playing a game in a hand-held virtual world together.

The feature stops children from falling into their own, isolated little game space; instead they fall into an isolated game space together. Children playing these games will regularly do nothing except play these games for hours on end. The immersion children can have in these game spaces can potentially be unending.

Online. Handheld. Toys. Television. Media outlets and game and toy manufacturers have figured out how to coordinate their efforts to constantly remind children about the virtual worlds, and everything their missing.

Inventions like the Nintendo Wii give some parents hope that young gamers will not become part of the fabric of their sofas. The huge breakthrough regarding the Wii is the incorporation of something called an accelerometer into the joystick users use to control the games. An accelerometer is the technological device that allows a piece of technology to know when it's moving in a terrestrial environment. It makes the iPhone screen re-orient itself from vertical to horizontal (think of the future mobile gaming applications for that.) With an infrared signal, and an accelerometer we finally have a wireless, truly usable game controller that doesn't require a person to memorize the functions of five or six buttons. So now a controller exists that also requires a player to do much more than twiddle his thumbs, s/he needs to move *off* the couch! The Wii is not a substitute for actually doing something, but it at least gets kids off the couch.

The Wii is also one of the premier drivers of new gaming customers. A recent industry article bemoaned how the sales of games for the Wii had not kept up with the expectations related to how many Wii gaming consoles sold (this was before Nintendo released Wii Fit). The reason for this was that the Wii, a video game that is coming closer to allowing us to easily interface with the computer (the game console), attracted many first time gamers, and the video games that came bundled with most of the Wiis – *Wii Play* and *Wii Sports* – were more than sufficient objects of entertainment. These two gaming titles (DVDs with a myriad of similarly themed games already loaded on them) have gaming options like baseball, bowling, ping-pong, and tennis. The Wii is helping further blur the line between what is real and what is virtual. A child – or even adult – in an environment with limited access to activities like tennis or golf can experience a reasonable facsimile of the activity at home on the TV.

In online worlds like World of Warcraft, for those who are already drawn to hard core video game culture, the dangers increase as the volume gets turned up on the experience, and gamers become increasingly immersed in their experiences. Many anecdotes exist around gamers – teens and adults – who spend so much time playing games like World of Warcraft that they fail out or simply drop out of school, or allow their real-world relationships languish. Websites thrive by providing support to hard core gamers for finding new ways to succeed in the games. Online support groups exist for women who have lost their husbands to games like World of Warcraft.

One student at Mt. Diablo High School in Concord, CA, David T. a self-professed gamer, mentioned "I've never had a chance to play World of Warcraft, because I have a bad feeling I might get addicted to it, so I'm trying to avoid that."

Step back to think about what the Wii has unleashed. Byron Reeves speaks to the effect that an immersive visual experience has on its participants:

> *There's a volume knob…that when you move to moving pictures to from still pictures, to color pictures to larger pictures, to high-definition pictures to virtual reality goggles to three dimensional pictures that allow you as a participant to move around in a physical space, the volume is just turned up, the needle just starts pointing even more toward fact than fiction [in the participant's experience of the image.]*

Not only do interactive video games have the visual element turning up the volume on players' experiences, but the untethered controller is now, slowly, allowing users to interact with a computer in an intuitive, physical way. As of 2008, that's a pretty novel invention.

So if the raised volume of user experience is where it is today, facilitating people to embark upon self-destructive behaviors like neglecting school, work, and relationships, what will happen as the physical experience expands, draws in people who were outside of the stereotypical gaming population, and leads people who are less and less inclined to understand the potential pitfalls of playing games for too long? Time will be the other variable that unveils the answer to this, but there must be an up-side to these interfaces that create a more terrestrial, "real" experience for gamers, right? After all what about the child in an underprivileged home who has no access to golf clubs, and a golf course? Isn't putting a Wii into the hands of that child, or adult, going to provide him or her opportunities he or she may potentially never have? Until the controllers provide a far different experience, some children are following a dangerous path; it is a fallacy of expertise, which actually is not limited to children.

Anecdotally, I find that children will believe they know how to play sports like tennis after playing round after round of it on the Wii. I recently had a child say to me, "I want to go play tennis. I know I'll be great. I beat my dad on the Wii." A boy I was coaching on a Little League baseball team this spring started the season with a great swing, and perfect timing – hitting the ball as consistently as anyone on the team. Slowly, however, his timing degraded and he couldn't hit a single ball. His father came to me and apologized, "it's the Wii," he said. The boy went on a 10-day vacation with his parents, and after he returned, after not touching his or any Wii for that time, his timing went back to what he originally had, and he's hitting the ball well again.

Many argue that the online tools of the Internet that make it so attractive are the very tools that are helping to undo our civilization, forget about the nuances of culture. An April 2007 article from *The Sunday Times* (UK) posted on its website, *TimesOnline.co.uk*, drew from a number of resources ranging from its own editors to a 2007 book by Andrew Keen, *The Cult of the Amateur*.

A quote from Keen's book goes so far as saying, "It's the cult of the child."

He continues, "The more you know, the less you know. It's all about digital narcissism, shameless self-promotion." (Appleyard, 2007)

In *Snow Crash* virtual world users were either totally immersed in real life or the virtual world. This science fiction scenario could not have predicted this near-term skill-set of today's digital natives. Today's digital culture is not limited to virtual worlds. Digital natives seamlessly move from one medium to another and in many anecdotal cases live in multiple spaces at once. A marketing director in a large communications hardware supply company gave me an anecdote regarding his own son, one that directly affected how he viewed marketing to one of their primary target audiences: teens. He came into his son's room one day to find him texting on his cell phone, IMing on one window of his computer, posting on MySpace in another window in the computer, listening to music on his iPod, and doing homework. Of course he could not effectively be doing all of them perfectly simultaneously, but he was seamlessly, and rapidly jumping from one medium to another.

As the *TimesOnline* article continues, the Internet is "creating a world in which everybody can talk — or, more commonly, shout — about themselves to everybody else. This is already changing politics, the record industry, print media, advertising and will, in time, change, perhaps to the point of destruction, almost everything else." (Appleyard, 2007)

This is a very damning position that is actually supported by people in academia.

An excerpt from an October 7, 2005 article by Scott Carlson, in the *Chronicle of Higher Education*, *The Net Generation Goes to College*, illustrated how professors at distinguished universities – professors who have taught for a while, and who have seen the shift in student behavior – see students today as being more ego-centric than years prior. The student skills lie more with speaking at others, than with listening and consuming information, processing that information, then entering into a discourse. (Carlson, 2005)

Is the freedom afforded by the web going to be the undoing of culture, and, eventually, society? It very well may be.

Back to the *TimesOnline* article it states,

> *Cultural continuity depends on arbitrary authority. There is no absolute justification for teaching children Shakespeare or maths; there is simply the necessity to teach them something that will place them in their world and show them the height of what we believe is the highest. But arbitrary authority is anathema to [digital society]. It is predicated not just on the wisdom of the crowd, but on its power. So, Wikipedia can be written and rewritten by everybody who uses it. Applying [previous arguments from within the article], this should mean it is the most accurate encyclopedia in the world...But, of course, it isn't. (Appleyard, 2007)*

Wikipedia and other sources of information consistently laud voices of authority.

A perfect illustration of this was of

> *...a prolific contributor [to Wikipedia] who was said to be a professor with degrees in theology and canon law, turned out to be Ryan Jordan, a 24-year-old college dropout from Kentucky. Jordan exploited the trust structure of the internet technology to pretend he was somebody else. (Appleyard, 2007)*

This, after all, is the greatest and worst element of the Internet. Anonymity. Looking at this anonymity, however, we can easily start to see that the costs of anonymity can quickly outweigh its benefits. As quoted above, "Cultural continuity depends on arbitrary authority." Anonymity allows one to skirt arbitrary authority. Using these definitions, a person can see the virtual landscape become one full of culture and totally devoid of culture.

And without a defined culture, what is our society? As Coye Cheshire comments, the development of culture is "a process, it's something that takes a lot time."

Chapter 3: Transition

"Just how radically is the Internet transforming the experience of childhood?" *Frontline* January 22, 2008.

Physical dangers. Financial juggernauts. Cultural vacuums. Coming back to what first started me down the path of this project I wonder how all of this is affecting kids. This is a terribly broad question, so allow me to narrow it to "how is it changing their developmental and behavioral patterns?" Are digital media exacerbating the problems of obesity? Do digital media create reasons why we should keep our children on shorter leashes, always under lock and key? Are digital media the reason why so many children today appear to have such short attention spans, and disrespect for authority? Is the Internet and exposure to digital media leading to a new use of physical human memory?

This is America, after all, and in America we beg science to find the root cause of large cultural and social issues, create a little blue pill for the issue, and dispense the pill through our HMOs and prescription plans, which allows us to continue on with our normal lives. What can the panacea be here? Well, before discovering the panacea, we need to uncover what is so different with kids today, what's causing the differences, and whether or not it actually needs curing.

When you find a child with a nose glued to a Nintendo DS, or a Microsoft X-Box, or even an innocuously named online kid's world like Club Penguin, you need to watch this child's behavior with the game; it's fascinating. Regardless of the platform (hand-held, TV, or computer) the child becomes totally immersed in the game. Totally.

Every parent with kids who play these games can attest to what some may consider the mesmerizing effects of these games, but I would conjecture that the children are not mesmerized at all; they are immersed in the games. They are a part of the games. Their experience within these games is as real to them as their going out to play a game of Wiffle Ball with friends.

Though many adults are struck with great fear when they see online, 3D immersive spaces like Second Life, cowering at the thought that "real" life is being usurped by an alternative, increasingly life-like virtual life, these adults should be more fearful of what already exists. The fear should not come from that which comes closer and closer to real life, but that which is obviously not real, which allows us – not just kids – to accept the suspension of disbelief, and become a part of these animated worlds.

These children are the puppeteers of their own digital existence. Driving the cars of the Mario Brothers. Flipping upside down on skateboards with Tony Hawk. Slaying Dragons in the virtual world DragonFable. All of these experiences are becoming part of children's collective unconsciousness, their own cultural capital and social currency that they negotiate and trade. If between two school-age friends, one friend is a fan of, and plays one game like Sonic the Hedgehog, while the other friend is obsessed with Lego Star Wars, an interesting struggle occurs, which is as indicative of the friendship the children have, as it is of the personalities the children are developing. One friend may not be as invested in his favorite game and easily acquiesce, deferring to the other friend regarding the discussion of what game is better. Both friends might be equally headstrong, and the negotiation over a favorite gaming platform can actually lead to a temporary dissolution of the friendship.

Negotiations over gaming platforms can occur at play-dates or on the school-yard at any time without gaming systems being anywhere nearby. The content of these games has become such a component of the school-age children's cultural fabric that imaginary games regularly occur involving the characters, powers, and scenarios presented in these games.

A child's imagination is powerful, and adults have regularly used a child's imagination to manipulate a child's development. Hundreds of years ago the Brothers Grimm scribed German folk tales that had been told for generations. These folk tales – Grimm's Fairy Tales – became important cultural touchstones in the developed Western world. Deconstruct these fairy tales, however, and you easily find the moral, ethical, and practical lessons the stories were imparting upon children.

Think of the classic story of Hansel and Grethel, of the intense fear that a story such as this would instill in children regarding straying from home, balanced with the trust in their inherent resourcefulness to find a means to escape. The innate aspect of children that these stories were leveraging in order to pass on these lessons was the power of a child's imagination, which allowed these fairy tales come to life. Of course the stories were hyperbole, but the use of hyperbole enhanced the imaginative experience of the children, further allowing the stories to facilitate passing along the messages parents wanted to teach their children.

Many parents and observers of today's children are quick to say that these young digital natives have no imagination. No, these children don't have imaginations like we did thirty years ago, but listen to their language; their imaginations are as rich as they ever were, it's just that their cultural context is something with which we are all not as familiar, and maybe more importantly, not as comfortable.

The cultural content that children use to construct their imaginary worlds no longer comes from fairy tales, or other children's books; it comes primarily from visual media. Movies from movie theatres, movies on DVD, television shows, and video games (handheld, console, and online), all add to the collective body of knowledge children use to build a sense of common interests, and shared experience.

Christian Renaud talks about the access to creative content in terms of his own daughters:

> It's almost too much. If you look at it from just an animal perspective, 'honey, you've got every kid's film ever made available to you all of the time, any piece of music that ... you've ever heard, that you can think of daddy has, and you can consume it anywhere. You can bring it with you in the iPod, you can listen to it in the car, or whatever.'

A child's imagination may be as rich today as it ever has been. What gets confusing is that children have so many more resources from which to draw to enhance that imagination. What is problematic is that the imaginations of today's children are manipulated by media that has limited interest in passing along moral and ethical values, and more interest in building strong merchandising brands. We want our children to be smarter than us, to excel beyond our means, to have better lives, and for those of us who are not digital natives it is nearly impossible to make the conceptual leap that a child immersed in the world of Sonic the Hedgehog is going to have the intellectual acumen to ever do well in anything except for the land of Sonic the Hedgehog.

This is not, however, a rationalization or justification for changed behaviors – if that's truly what has happened with children –

merely an observation that the content of children's imaginations has changed, not the children's ability to use their imaginations. Another observation is that these kids who are so immersed in these digital environments are losing certain basic skills of manual manipulation.

Kindergarten teachers note that many of the children coming to school now don't have the simple skills of using scissors, something that almost all children 20 years ago would have had.

A kindergarten teacher at our sons' elementary school, with 25 years of experience noted "most parents don't let their kids have scissors, don't let them tear up things. There are very little activities that help their kids develop their fine motor, and consequently when they come into Kindergarten, and we have to put that pencil quickly into their hands, it's very challenging."

She has also observed that few of the children have the inclination to read on their own and that few of the parents are reading to their children at home, instead allowing the games to be an alternative to occupying the children. One could conclude that for some children the amounts to which parents try to protect leads to some of the physical and behavioral differences in today's children.

I helped out my son's first grade Physical Education classes, which made for an interesting observation pool. It was simple to see the children who regularly got outside to play, and to see those children who regularly played video games, or watched TV instead. Compound this growing propensity for sedentary lifestyles with the use of prepared and processed foods in harried households where more often than not both parents work, if and when two parents are even in the household, and it's easy to see how digital media are a significant component to rising childhood obesity rates.

With a tool like a TV, or a Nintendo DS, or Sony Playstation, or access to online games, a child's resources for self-play expand, but the resources are also limited to these tools. Before the video game era even television was a limited option for children. For example, in 1975, a child coming home from elementary school in the Boston metropolitan area, one of the top five media markets of the time, would have had two channels of children's television programming available between the hours of 4PM and 6PM. After 6PM, television programming would revert to adult programming, and a child's ownership of the television was over.

Conversely, in 2008, a child coming home from elementary school, who has access to cable television, has access to 18 channels of children's programming – or what is defined as programming targeted to children – and most of these channels are available to children 24 hours a day. As of 2007, out of the 112.2 million US households, 58%, or almost 65 million households subscribed to at least basic cable services. (NCTA, 2008) The US Census Bureau statistics indicate there is an average of .94 children under 18 in each household, which means that just fewer than 61 million American children have access to this programming. (USA, 2007)

And this is just about cable television. Video games are always available. Cable television provides on-demand programming, which was an impossible service to offer before the integration of computers into the distribution of television signals. DVDs provide even greater access to media exposure. Children have become so accustomed to on-demand entertainment through an almost endless pool of visual and interactive media, that the attraction to go outside and play Wiffle Ball, or ride a bike around the neighborhood seems almost pedestrian. Couple this access to media with parents who are spread too thin attempting to manage cash-strapped households, and you see the dangerous possibility of children never making it outside to play.

The same media outlets that entertain our children entertain and inform the children's parents. If you are bombarded by recurring news stories of child abduction and the dangers facing today's youth then you can see why most parents today rarely let their children explore past the boundaries of their yards. Many parents won't even let their children play in the front yards due the fear that many media outlets have instilled in parents regarding the vulnerability of today's children.

danah boyd provides great historical perspective about kids today, specifically teens,

> *...they're the most locked down group you have. There's ...unbelievable research done by a man named William Byrd, and he works for some Byrd Society in the UK, and he was obsessed with nature and the different generations' access to nature, and was looking at how people have less mobility. Teenagers have less mobility now than ever before. And a newspaper printing up his research in the UK decided to build him a map, and they*

*used four generations of eight-year-olds in Sheffield.
The grandfather could go six miles to the local pond,
and then the father could go...downtown to the candy
shop, the person being interviewed could go around the
neighborhood, which is basically the generation of
young and middle aged parents today, and the kid
couldn't leave the yard.*

But herein lays the double-edged sword of digital media use in children. While digital media from handheld games to online worlds can provide parents a brief respite while cooking dinner or cleaning the house, and give children relegated to the front yards something to do, these digital media also require a significant amount of supervision. And while our adult needs focus on household chores and earning incomes, we restrict the movement of our kids, which encourages youngsters to convene online.

Kate K., an elementary school student in Walnut Creek, CA, said while sitting next to her friend Christine S., "She's on [the online game] and I go with her."

Christine S. added, "We're friends there."

What Christine S. is pointing out is how in most of these online worlds geared for kids – and even ones geared towards adults – you cannot interact with someone unless you ask them to be your friend, and they accept the invitation.

Kate K. continued, "We communicate on the phone. We can just talk to each other while we're playing."

These girls are third graders. The older elementary school kids are even more involved. When asked about making friends on sites like Club Penguin or Webkinz, a group of fourth and fifth graders blurted out, pointing to each other "I have you on my friends list."

"You're on my list on Webkinz!"

"All my friends [from school] are on my friends list on Club Penguin."

When harried from a long day, and I, the parent, need to cook dinner, or finish last minute work for a job, I can feel better about myself by encouraging my sons to play with their friends online. It's a lot less intrusive for me than driving them across town to a friend's house.

So when I'm busy in the afternoon cleaning up after the kids and getting ready for dinner, and I've pushed a handheld video game into the hands of my children to baby sit them for the two hours I need before dinner, the question arises, "How violent is the handheld game? What kind of language is my child using when playing something as innocuous as Lego Star Wars?"

I recently found our Kindergartner mumbling "die, die, die" while deeply immersed in a "Mini-game" in Lego Star Wars. Lego Star Wars is a brilliant game for kids, which has versions available for every major gaming platform from a PC, to Nintendo DS, to the Wii. The version our youngest was playing was on his DS, and in this DS game the "mini-games" are short tangents from the narrative of the larger game that kids can take to earn more coins in the game, with which they can purchase access to different characters and other game-specific items. The coins are really little Lego buttons; everything in the game is something that could be built in the real world with real Legos. The mini-game my son was playing in this case was some shoot-em-up gallery where the only object of the game was shooting as many other characters as possible. His game – and all other video games – was quickly replaced with other non-digital games and activities.

I like to think that the process of removing the games for a short while sent a message that behavior such as saying "die, die, die" is not acceptable, but I know I'm fooling myself, and I am somewhat fearful of what the true repercussions of his exposure to these games at such a young age is going to have in his early teen years.

But why do I acquiesce and not play the role of parent more sternly? I remember very distinctly as a child what it was like to meet those few kids whose parents did not allow them to watch TV. They verged on being social outcasts. There were no cultural touchstones on which to connect. Even if the kids were really nice, the relationships would only go so far because the points of reference were so far apart. Like it or not, digital media of all sorts are the TV of this younger generation. To totally withhold access to the media is to stunt the children socially, which is a significant element of development because, after all, we are social creatures.

Responsible parenting in this new environment means monitoring media use: what and when, and for how long. Just as my parents did not let me stay up late to watch *Hill Street Blues* when it first aired because it was not only on too late for me, but its content was still a bit too mature for me, we parents today need to monitor what media children consume today. This, however, is a much harder task today. As indicated before, when I was a kid in 1975 there were eight TV stations that included fuzzy UHF channels. Watching TV sometimes took the skills of a contortionist, and sculptor of aluminum foil. If reception on the television was bad, finding the sweet spot for the best reception with the "rabbit ears" antenna was often a fitful pursuit.

On top of the media options discussed before, there are websites with chatrooms, and simple Google searches that, even with properly installed parental controls, require oversight. Given all of these threats it's no wonder that many families place the only computer that the children have access to in a kitchen or hallway.

Short of not giving children access to any form of digital media, parents have no other recourse than to be vigilant in understanding and monitoring the digital landscape children are traversing. The added difficulty in this vigilance, however, is the mistrust connoted to children, particularly as children reach the Tween and Teen years where it's their job to seek autonomy and to stretch boundaries. Practically every child will be insulted by a parent's efforts to monitor the child's digital activity, which becomes increasingly problematic when the child is old enough to have a cell phone. How does a parent track a child's texting history without making the child feel his or her personal space is being violated, or that he or she is not trusted?

If the foundation of rules for digital media use has not already been established, inserting one's self into a child's digital landscape is, probably, impossible. With the newer, younger digital natives, setting a pattern early on regarding the use of digital technology will hopefully make for a smoother transition into the problematic teen years.

But what of the patterns already established by the older digital natives: the ones in college and in high school, up to whom younger kids look? The *Chronicle of Higher Education* article referred to earlier talks about this first wave of students reaching college: the Millenials. The article asks if the educational paradigm as we have known it for years is outmoded. Given the multi-tasking skills of this new generation, is the Socratic Method of old no longer relevant?

There are many educators who think that the educational paradigm does need to shift.

The ego-centrism, and need the students have to be heard increasingly reduces the respect of, and importance of the professors. Instead of being esteemed for their experience and knowledge, the article claims that many students view the professors as irrelevant talking heads.

The danger in this goes far beyond the deconstruction of the current mode of how education is conducted in America. No longer are universities able to teach students how to sit and contemplate, to reflect, and to deeply analyze. Analysis is an effective Google search – or whatever Internet search tool usurps Google in the future. (Carlson, 2005)

Given this shift in how some see children learning, does this mean we are approaching the threshold of the Singularity? The notion of the Singularity, as defined by Ray Kurzweil in his 2005 book *The Singularity is Near: When Humans Transcend Biology*, is that in the near(ing) future, a technological innovation will come along that will so influence society that it will inalterably change human culture: how we interact with the world around us, how we process information, and how we relate to each other. Listen to comments from Christian Renaud, or Jamais Cascio and you will see how we may be closer to the Singularity than you think.

Christian comments,

> *It's tricky, because I have a really, really, really, strong concept of free will. To quote Heinlin, I make Ayn Rand look like a socialist. So the libertarian strains in me reject this hypothesis. The technologist in me, which is also dominant, sees that human beings, given the opportunity to use technology – [after all] we're omnivores and not herbivores – we adapt as a species, we use tools, and we become dependent on tools, and if you ask the average human now to operate without any tools at all, if you ask them to go to the farm and actually subsist without tools, they're incapable, right? Look at television shows that put people on islands. [We're] incapable of ... [living without tools] in this modern age. So I think we're going to become*

increasingly dependent – to the point of overdependence depending on your perspective – on these tools. And at a certain point...At what point does it stop being entirely human – using tools – and the two of them working together? Not a cyborg per se...It's not a compliment any more, it's an essential, and I can't be functional without it. Life memory aids, and prosthetics, that's where the line starts to get fuzzy. [There's] the work being done at MIT in smart prosthetics, memory augmentation, people that are being treated for Alzheimer's using technology and so forth, that is the beginning of, and it always starts in a vertical market, what's going to become commonplace, and pretty soon when your child doesn't have the memory augmentation implant they're not going to be as competitive against the other kids who don't have parents who have the philosophical qualms against memory implants, and they will have total recall and your child will not. So there is some degree of inevitability.

Jamais Cascio has more questions than comments.

What we really need to be looking at, and where I think the really interesting stories get told, are around how are people changing? Are relationships changing, whether it's technologically enabled or otherwise? How are our capacities to communicate and understand each other changing? What are the implications of these seemingly simple devices?

Technology's influences are far-reaching and not just limited to texting, chat rooms, or video games; it's changed everything. There's is no escaping the influence, and it's just going to get greater. But are the effects all bad?

Section 2: Perfection!

Chapter 4: Utopia

"**Utopia** An ideally perfect place, especially in its social, political, and moral aspects." – *The American Heritage® Dictionary of the English Language*, Fourth Edition.

"We're in the very beginning stages of what this technology is about," says David Fleck, who as of January 2008 was with GoPets, another virtual world targeted to pre-teens, teens, and older; some say it's a virtual dog park. David is a virtual worlds marketing executive who had been with Linden Labs – the creators of Second Life – before going to GoPets. Listen to David for a while, and you realize his passion for both virtual worlds, and technology. His passion, however, comes with fairly healthy doses of perspective regarding who drives innovation in virtual worlds.

David's comment about this being the beginning stages of the real explosion of Computer Mediated Communication, particularly in virtual worlds, illustrate why most virtual worlds involve gaming, which dovetails with Byron Reeve's comments about the history of media,

> ...gamers and the games themselves often take the lead here [in the development of technology,] and that's been true in the history of media, and not just games, but entertainment uses of media often are the first experiences and then we figure out serious things to do with that entertainment, in large part because the entertainment is important for the serious application. So I actually think that the prospects for games to...for that background to be relatively more silent, as serious uses come into play in these virtual worlds is really large. People even now are starting to have serious relationships; they're starting to not only laugh, but even cry, to organize politically, to experience religion, to do all kinds of things in these virtual worlds that really take it out of the pure game play origin.

David Fleck's addition to Byron's comments is more of a cultural criticism,

> ...at these earliest stages [of virtual worlds] quite often people look at them as being more entertainment-based than anything, and as a result it's kind of a game mentality that people bring to the table when they look at the products, or the technologies that

are out there.

Of course, then, most parents of digital natives are going to look at their children's activities and consider them superfluous, and inconsequential, except for when parents are actively excluded by a child's texting a friend sitting right next to him or her.

Curious, however, is how children know to model this behavior. Language may be innate, but not thumbs on a keypad. Children mirror others: parents, siblings, and media. With our own addictions to Blackberries and iPhones, it's no wonder kids fall into their texting worlds as soon as they are able, and no wonder that children as young as eight years old pester their parents, wondering when they will be old enough to have their own cell phones.

But even if you view all current technological innovations as more game than "serious" endeavor, even a cursory examination shows that not all games are bastions of physical and moral destruction. Most games start with the simple premise of challenging the mind, of allowing players to pit their skills against each other, of creating an environment that is fun. Here's where the discrepancy begins, but that discrepancy is one of semantics, not of right and wrong.

One of my favorite professors, Maria Koundura from Emerson College, once gave me a wonderful illustration for describing semantic differences, and how easily the simple words we use can become confused. Take a handful of friends in a room, and give each of them a pencil and piece of paper. Regardless of drawing ability, have each of them draw a picture of what they see as a cup – no further definition – just a cup. And this drawing shouldn't take a lot of time; we're not looking for perfect perspective, or shading, or artistic nuance. I did this with a class of twelve students on more than one occasion, and each time out of twelve students you would find at least eight different kinds of cups.

Let's now apply this exercise to the word "game."

The American Heritage Dictionary lists ten separate definitions of the word game. Including further details of the definitions under each of the primary definitions, there are fifty definitions of the word "game." This includes the use of the word game in a hunting context, but that is just as easily applicable here.

You might ask how this applies to the Utopia of the digital

universe – of Computer Mediated Communication – but it is very simple. With search engines, and the imagination of the human mind, users of technology can now easily find spaces where his or her cup looks exactly like everyone else's cup.

Why is this good? Communication, good communication, is predicated on each of the participants using the same lexicon. Without each participant agreeing upon the library of words s/he is using, confusion becomes inevitable. As Aaron P., another high school student I interviewed at Mt. Diablo High School, in Concord, CA,

> *...when people see you, a stereotype kind of plays in, you might act a certain way, you might act or like certain things that a lot of people in your area don't really like, or don't really like about that,...or will put you down for it. When you're on the Internet, it's easy to find a whole room, a whole environment of people who have the same feelings, the same likes as you, and you fit in really easily. You might have that locked down feeling in your house, in your world...Online it's a whole new world for you, you actually fit in, you can be yourself, and not have to worry about being knocked down for it.*

This is Computer Mediated Communication: social networking, texting, games, blogging, and more. They are all reflections of community – where people tend to hang out – which builds upon the notion of shared interests, shared values, and shared lexicons: the libraries of words we use with our peers.

Social networking sites have grown up alongside online video games. As soon as there was a commercial Internet available, a region called the Usenet, which still exists today, emerged. Usenet has been extensively researched in academic circles; it is the rawest form of Computer Mediated Communication on the Web, one which formed without guidance, without corporate facilitation, that grew solely out of the desires of the users. Usenet spaces had wacky Internet names like alt.binary.motorcylces. In these places people who liked motorcycles, or kittens, or art deco graphic design came to ask questions, reply to questions, to find others who liked the same things they did, and show others that they liked the same things as the others in that little corner of the Usenet, which was just one little corner of the growing Internet. What's significant about the explosion of the Usenet was that it was totally text based. A user would write a comment and just "post" it, meaning that it would then appear as a small paragraph, or paragraphs, in its specific Usenet page categorized by interest. That posting would then elicit responses from other users, or not. Sometimes the threads of responses would take on entire lives of their own. Sometimes a post would receive one, two, or even no responses. It was, and still is a large free form bulletin board that allows unlimited graffiti and serious discourse all in the same place.

The Usenet was not, and is not, just relegated to simple interest categories – serious discussions unfold here. As danah boyd illustrated,

> *...in old Usenet, the Palestinian and Israeli conflict actually was taking place in two different forums simultaneously: one was a Jewish, Israeli forum, and one was a broader Arab forum. The result is that each of these conversations assumed the context they were taking, that they don't have to talk about all of this, and we understand the bend that I'm coming from.*

An important note here involves the currency that was used on the "old" Usenet. When you communicate with someone you are expecting they are hearing your words a certain way. You expect some words to be inspiring, and some to be inflammatory. When, however, an unexpected audience hears your words, you have no idea how those words are going to be received; you have no point of reference for how they will interpret what you've said.

danah illustrates this with by continuing her with her anecdote about the old Usenet.

> *Search [like Google or Yahoo] puts them in the same world. [PAUSE] They don't. And all of a sudden [you mash together two lexicons], and they don't match. It's a struggle for how are we dealing with these multiple contexts, and we have to move between them. We don't know what the correct response[s] to these things are.*

Human need for communication, and shared interests followed into the online gaming community. Sibley Verbeck is a consultant in the space of Computer Mediated Communication focusing on virtual worlds. He is the CEO of a firm called Electric Sheep. In an interview with him he noted, "virtual worlds have done a terrible job at social networking."

I make the leap to Sibley's comment because except for undirected sites like Second Life, where there is no purpose to the site except to create a life for one's avatar, essentially to create a digital life for one's self, or for corporations to create some form of virtual space in which anyone can visit at any time, most other virtual worlds, like World of Warcraft, have very specific purposes. These are online worlds with very rich graphics, and very distinct purposes all related to the games the sites represent. Without being designed for it, however, these virtual worlds have become social networking sites as well. Guild members become close friends. The shared purpose involved in the game becomes a strong social bond, much stronger than anything found in a website specifically geared to professional social networking like LinkedIn.com.

Coye Cheshire says,

> ...shared experience and the ability to sort of talk about and share things,... being able to share experience, talk about, exchange our stories about the same kind of thing, that's a way in which we build all kinds of positive social outcomes like group solidarity, and all kinds of things that are really important for a community or a society, even more broadly.

Many people also argue that the Computer Mediated Communication spaces like the Internet are ruining culture, as seen from the prior reference to Bryan Appleyard's column in **The Sunday Times Online** article from April 22, 2007 that uses Andrew Keen's book **The Cult of the Amateur: How Today's Internet Is Killing Our Culture** as an example that the populist problems inherent in the Web may easily outstrip the benefits of the Web. Keen, as referenced by Appleyard's column, and in an a June 5, 2007 **Wired Blog Network** post by Jeff Howe, does not believe that the egalitarian prospect of the

Web will move society and culture forward, instead Keen is a grand proponent of the old methods of bringing creative content to the masses.

He says in his interview with Howe,

> *I don't want the crowd to tell me what's worth watching. I want a movie critic to tell me that. I don't want the crowd to tell me where to eat, because I don't trust them to know. Give me the old gatekeepers any day. (Howe, 2007)*

Keen is not alone. Art critics, movie studio moguls, music industry executives, and literature publishers all bemoan the ineffectual use of Digital Rights Management (discussed earlier), and how the digitization of creative arts has lead to the rampant piracy of movies, music, and more.

For one hundred years the executives of these industries have maintained a well established system of gate keepers insuring that their vision of good art makes it to the marketplace. After all, the commodification of art is a funny business. What is good? What is bad? Who determines what is good and what is bad? Many fortunes have been made and lost on one executive's decision regarding what will be popular and what will not.

As we all know, the Internet has changed that. Even when the industry executives had no desire to make their arts digitally available, users of the Internet found ways to make it digital, and to distribute it. The music industry was the first to really feel the effects of this, as with the proliferation of Compact Discs before the Internet revolution, music was already poised to be rampantly copied and distributed. Music on Compact Discs is already in a digital format: something a computer can read, manage, save, and pass on to another computer. The Internet is a highway system for digital information to make it from one computer to another. Before the advent of either the Compact Disc, or the Internet, people were already well adept at reapportioning music. First with reel-to-reel tape decks, and then with the easier to use, and less expensive cassette tape deck, teenagers and adults created mix tapes of their favorite music, or simply transferred an LP from the turntable onto a cassette tape to play in the car, or to give to a friend.

At first the music industry fought this form of piracy, but they found that the proliferation of the cassette tapes actually helped bolster their Industry, broadening exposure of their artists. The time involved in creating the tapes, coupled with the longevity of the tapes lead most people who received the tapes to purchase their own copies of the LPs: a medium that did not degrade the way an often played cassette tape did.

When the Internet and Compact Discs reduced the time to mere minutes for creating music mixes, or copying entire albums, change in the Industry was inevitable. File sharing systems like Napster quickly arose, which allowed thousands of computers to see each other's libraries of music files, and to quickly copy and share those music files.

The process was still increasing the exposure that the Industry's artists were receiving, but no one in the Industry was seeing the money any longer. Music Industry executives quickly cried foul and the US Department of Justice stepped in to make example of the most prolific music file copiers, and to shut down the web sites that facilitated the practice. (A&M Records, Inc. v. Napster, Inc., 2001)

The movie and literature publishing industries were able to avoid this rapid proliferation of piracy for a few years as the technology had not yet caught up to either transfer the huge size of movie files, or quickly transfer printed pages into a digital format – the large push by companies like Google to digitize the great works of literature had not yet begun. A national announcement from Google, as referenced from *Information Today* in October 6, 2004, spoke about Google's Google Print program, which started October of that year. The article described how Google's program wanted to digitize any materials that publishers were willing to provide. (Quint, 2004)

This move by Google continued along the trend that, once the technology had caught up, would force all of the gatekeepers of the commodofied art world to rethink how they made money; this was the threshold of Utopia in creative content.

Industry executives in creative media talk about the threat of how China and other Asian powers disregard Western copyrights, a further burden to the piracy of content. While US courts could convict US nationals, it had no jurisdiction over other countries, and countries like China provide a lot of lip service over the protection of US copyrights. The irony is thick, however, since most US literary publishing powerhouses made their initial fortunes on the backs of Western European copyrights that they disregarded at the turn of the last century. Jamais recounted the following history and put it into context regarding the near future, augmented reality, and the growing practice of "life logging," a term defined in the ***Metaverse Roadmap Summary***.

To quickly define life logging, it is the process of documenting (textually, or through audio or video recordings) the entirety of one's life. Internet spaces like Twitter.com are already allowing people to textually record and share with others brief snippets of what users are doing at that very moment...like little birds twittering about.

On Twitter a user will type a 200 character phrase describing what s/he is doing at that moment, which is then available for anyone to see. Potentially tens of thousands of people could immediately read that Twitter, but usually its relegated to a small handful of close acquaintances, which raises all sorts of other interesting questions not only about intellectual property, but around privacy, and relationship dynamics.

Augmented Reality has the capability of being a part of life logging, but is something separate. By using computers and computer networks to their fullest extent, finding ways to relate all of the information that is stored on all of the computers around the world, tools will appear that weave that information seamlessly into our everyday lives. Augmented Reality (AR) integrates where you physically stand with a body of knowledge greater than you can realistically access on your own, feeding you information on the people around you, the stores near you, the restaurants close by that your friends like to frequent. Ideally, AR also comes to you visually through some means: specialized goggles, an implant, or contact lenses. In a recent blog entry Christian Renaud discussed the recent advances with tools like the new 3G iPhone, that not only has GPS (Global Positioning) features that allow it to know exactly where on earth the phone is, but integrates that information unique to the owner of the phone,...if the owner of the phone tells the phone what is unique about him (or her). He intimates that we may already be on the cusp of AR.

So, what is the difference between the traditional definitions of Augmented Reality and a handheld device [like the recent 3G release of the iPhone] that has, as we used to say in the Emergent Collaboration team at Cisco, "The three C's: Contacts, Content and Context"? Probably just the heads-up-display, although I am sure some savvy Apple developer could make my newest VGA goggles do a pretty good imitation if there wasn't a police officer somewhere around to ticket me. (Renaud, 2008)

As AR starts to become more available, how people use AR, and what AR will expose people to, is going to open up all sorts of ethical and practical questions regarding intellectual property and privacy. Ownership of information that is easily capturable through digital means (microphones, or video or still camera capture) will fall into a bucket that danah boyd, Coye Cheshire, and Jamais Cascio all call the new "publics." The old definitions of public and private will no longer apply.

Regarding specifically intellectual property, Jamais says,

Let's say I have contact lenses on – the augmented reality contact lenses – that don't just display things for me, but record the world, because, y'know, we're going to want to do that – we're going to want to record things that happen to us – we're doing it now, and I go to the movies, what happens? Is there some sort of signal that automatically shuts off my AR [(Augmented Reality)] system so that I can't record the movie? What happens if I'm walking through a building and there's music playing that gets recorded? I'm making a digital copy of a piece of music. Do I suddenly get dinged a couple of cents from my bank account just because I happen to pass by something? There are these really sticky questions around ownership of ideas and media in a world where we have the capacity to surreptitiously or at least in an almost undetectable way, absorb and record and document the world. Almost inadvertently.

But the Utopia is not coming out of the sticky questions

regarding what are essentially concerns that stem from Digital Rights Management discussed earlier; it is blossoming from growth of the increasingly egalitarian Internet.

Am April 27, 2008 *NY Times Magazine* article (Heffernan, 2008) concluded with an anecdote how a poster to the website Flickr uploaded a seminal piece of photographic history as an experiment. While many elements of the article detail how the art of photography is changing, it is also trying to understand where the new barometer of taste stands. Within the article it speaks about how Flickr, a website where anyone can post any digital photograph, as long as they own the copyright of that photograph, meaning that the poster of the photo was the person responsible for either taking photo or now owned the rights to the photo. The article follows the artistic development of Rebekka Guoleifsdottir, a one of the most popular posters of images on Flickr. Interestingly, just like in real life, because Rebekka is a popular poster of images, she is also a driver of taste and style on the site.

The anecdote in the article is about a poster who uploaded a copy of a famous photograph of a person on a bicycle passing a spiral staircase. In high-art circles, this photograph by Henri Cartier-Bresson is considered the pinnacle of the art form. On Flikr, however, the photo was panned: railed upon as being the work of an inexperienced hack.

So what does this say about style and taste? Is it merely a reflection of shifting norms, and changing tastes, or is it the mark of the fall to ruin of our culture? (Heffernan, 2008)

This begs us to ask, what is good? Where and who are the gatekeepers of taste?

The vox populi becomes the gatekeeper of taste, something that should strike more fear into the hearts of creative content industry executives than any issues over people stealing already created works. For anyone who knows the workings of the creative industry, you know that a very small percentage of artists ever make significant money from their creations.

While they were not the first band to do this, Radiohead stepped into the growing arena of self-publishing their music. Their fan base allowed them to pocket as much money as they would have if they had allowed a music publishing label to distribute their music, and yet they had allowed their fans to only pay as much as they wanted to. Of course, some fans took the music for free, while others paid as much as they would have if they had purchased the music through a music store, or downloaded it from iTunes.

From a **Listening Post** posting on Wired.com by Eliot Van Buskirk,

> *According to ComScore's numbers, 1.2 million people worldwide visited inrainbows.com from October 1 to October 29, and 38 percent of them paid for 'In Rainbows,' despite having been presented with the option to download the album for free. Among the 456K people worldwide that ComScore says paid, the average price was exactly $6. If ComScore's sample is 100 percent accurate, that would mean Radiohead netted $2,736,000 in digital sales. US users (40 percent) were slightly more likely to pay for the album than users in other countries (36 percent). However, when they did pay, they paid nearly twice ($8.05) what their international counterparts did ($4.64). (Buskirk, 2007)*

Many Industry executives branded this as a failure, but given the net gain for the band, and the band's total control over their product, for whom was this experiment a failure?

The work of thousands of visual and musical artists, and tens of thousands of writers has seen the light of day, where before the Internet, access to the works of these people would have remained forever obscure. After all, access to the gatekeepers of taste has also been as important as sheer talent, as talent alone has not always guaranteed entrance into the hallowed halls of commodofied art. Without the advances of the last decade, I would not have been able to pursue this project the way I have.

Fifteen years ago I would have been limited to the use of a 16mm Bolex camera shooting on film that I would have to find some means to edit. The sound would have to be synced to the film. The process of writing this companion book would have been just as easy back then as it has been today, except I would have had to capture all of the interviews on an audio cassette recorder or similar medium, which would have been less easy to manage. I can see visual icons of the interview clips I have taken with my video camera, which, after fourteen months of manipulating them, are now icons that also remind me where to find each audio clip I want to reference. And the amount of research I have been able to do online would have taken me months more worth of time before the proliferation of the Internet. This book and its companion film are an indirect product of these technological advances. Advancements in technological tools have dramatically lowered the barrier to entry for people who want to create creative material for, and more importantly distribute it to the masses. An egalitarian Internet is not just about opinion; it's about access and creation.

Having an online space, however, does not guarantee that people will find the works of talented artists. Ironically enough, the art of shameless self-promotion still bolsters the career of aspiring artists. No one will go to a website if no one knows it's there, the same way no one will go to an artist's open studio if no one knows the studio exists.

Websites like Digg.com, or Mixx.com, which is newer and currently hipper than Digg.com, allow people to "tag" news articles and other web pages to let others who browse or use Mixx.com, or Digg.com know that something is cool or worth reading. Combine these spaces with social networking spaces, YouTube, and cell phones, and suddenly we have a new Utopia for the creation and consumption of art. No longer are the gatekeepers the sole determiners of taste. The vox populi will determine what is popular, while the constructors of taste, those on the fringe – the avant-garde – who have always been the subtle constructors of taste, will still play their role. The age-old mechanisms differentiating art and high-art will continue. What will be different is how the power to determine taste will no longer be in the hands of the few.

The Utopia of Computer Mediated Communication sits with the masses, as well as individuals. For individual digital citizens, one aspect of this new Utopia lies with its anonymity. While many see the anonymity of a digital frontier as a mask for potentially harmful activities, that same veil of anonymity is something that can be very beneficial to those who have always been looked upon poorly in their real-world physical visage.

In the digital world no one knows what you look like, or about any of your socially unacceptable ticks. In a digital world you can create the appearance of the person you always wanted to be. If you are a man who always has felt he would be better suited as a woman, you can take on that persona in a virtual world, and no one can tell that you are anything but. In a digital world you can become your true, authentic self, a goal for which most major religions want you to strive.

In how many spaces in our terrestrial existence can we simply shed the trappings of the person others have come to see us as, and reinvent ourselves in the image of how we see ourselves? Let's look at the second definition of "avatar" from the American Heritage Dictionary, "An embodiment, as of a quality or concept; an archetype: the very avatar of cunning." Using this definition you can see how the use of the word avatar as a digital representation of something in a virtual world is curiously accurate. As we saw in Chapter 2, an avatar in a virtual world is so much more than an animated character for which you are the puppeteer. How you control it defines your presence in a virtual world – your qualities, your personality – at least the personality that you are consciously deciding to present to others in this virtual world.

Except for spending an exorbitant amount of money to alter our appearance with a new hair cut or more extreme cosmetic surgery procedures, or purchase new clothing, or even a move to a new town, our terrestrial options for dictating how others see us are limited. In the digital world our options are limited only by our imagination. If you've always wanted to be a troll you can be one. In a place like World of Warcraft you can be a troll and surround yourself with others who accept trolls, and work together toward common goals, creating community through shared experiences, and shared struggles.

In a place like Second Life people won't think twice about seeing a troll, or a woman with wings, or cat ears, or be surprised to find out that the insanely attractive and flirtatious female avatar is really a male stock broker from New York City (as a potential hyperbole.) I've always wanted to be taller, and in a virtual world there is nothing stopping me from being a few inches taller. If I saw this as a greater manifestation of my true inner self, I could easily make my Second Life avatar, for example, 6 feet 4 inches tall, muscular with a chiseled jaw, and perfect hair. What I have done, however, is create an avatar that looks as close to what I think I look like in real life using my skills and the tools Second Life provides for creating avatars. In Second Life my avatar is slightly under average height, which is what I am, but

I must admit that the avatar is a bit more muscular than I am in my terrestrial life.

So this is the slight tweak that I have created in portraying what I see as my more authentic self – the self I would like to be, without having to spend ten hours in the gym every week while still enjoying Raisinettes as often as I would like.

Byron Reeves speaks to this phenomenon,

> *If you made someone that was five feet tall [grow to] eight feet tall that would be obvious, but you can do things like clean up faces. You can morph a visual of a face to match someone you are talking to. For example, if I took my face – a digital image of my face – morphed that thirty percent with your face I would still look like me, but you would actually like me better. You would be more persuaded by what I'm saying; we know that from studies. So even small changes are going to be important, and when images are digital there's a lot of opportunity for those changes to really be done pretty easily.*

After all, first impressions and attractive imagery are powerful tools in our society. Look at the successful sales forces in almost any industry; you don't find many unattractive people. Sales executives understand this. Imagine someone uncomfortable with, or cognizant of the limitations that would be placed in front of him – consciously and unconsciously – suffered from something like Asberger's Syndrome where the person knew they had an inability to pick up on visual and verbal social cues from others. Interfacing with people through a virtual world – through an avatar – eliminates many social stigmas.

Because someone does not fall within this group of people who create avatars that look mostly like their terrestrial visages does not mean that the outliers are up to no good. Think of the advantages that exist for someone who doesn't physically fall within demographic and aesthetic norms. Imagine someone who has a hard time grappling with a recent infirmary, like paralysis from an accident. This person would obviously pick up on the visual and social cues from others ranging from empathy to fear, which can lead to a habit of cowering from public situations.

This infirmed person can create a perfectly able bodied avatar in a virtual world and return to feeling those endorphin rushes s/he felt before the accident, shedding the perceived stigma, or just feeling liberated from the wheel chair for a few moments in time. In a virtual world you get to choose the image to which others have their first impression, instead of having to cope with the image you are born with, or have been handed since birth.

What could be better than a place where you get to choose every element of the person you set forth for others to see? danah boyd addresses this on a macro level. She says,

> *Hannah Arendt, a German scholar from 50 years ago, was looking at what makes public so valuable, and she talks about how the public is where we make things real. We make things real by enacting them, by engaging them, by performing them, and in that performance we test it, and we see the response to it – and this is Erving Goffman's notion of Impression Management – I'll flit my feathers, and I'll look like this thing, and you give me this response, and that wasn't quite what I wanted, so here's another performance, and you give me your response, and that's better but not quite what I want. And we go back and forth, and this is this question about identity. Is identity real? We're always performing, and acting identity in order to see if...our performance matches with the responses we want. And that's the whole game, because identity is a game of impression management; it's about saying 'I have a sense of who I am, and I want to convey it to you' but depending on your [response], like I'm completely nuts, I know whether or not I'm succeeding. And we have to work that through, and we work it through with different audiences at different times.*

In the near future, and even to a limited degree today in places like Second Life we see what Byron Reeves notes,

> *...in these virtual worlds, I have a representation of myself that I'm basically the puppeteer for, so I*

basically have the strings attached to that…Is that thing on the screen there, is that me? Is it a mini-me? Is it actually a third person, or a second person separate from my body, or a third person in some conversation with another individual?…We know that I'm very attached to it, I care about it more than somebody else's character or avatar.

All of this reinvention is currently occurring in a game space. Any industry conference on virtual worlds today is laden with former and current gaming executives. What is not lost on these business people is the power of 3D representations. As Sibley Verbeck said, "it's an easy story to tell."

Chapter 5: Retooling Business

"The internet [sic] has been a revolutionary technology, and the speed by which it has transformed business has no real precedent in history."
– **BBC News Online**, August 3, 2006.

As development of **IMHO** progressed I wanted to talk to the people creating these new worlds. I started by going to trade shows and conferences where I learned much about the commerce surrounding making money creating these worlds, and began to witness how much influence the executives from former gaming companies were having on the development of these worlds.

The Internet has transformed business. This next phase of the Internet, what many have dubbed Web 2.0 is, as Christian Renaud has put it, "the ultimate [online] convergence of social networking, gaming, and the enterprise."

What does this mean? As mentioned before, new communication technologies often start their introduction to society through games. MMOs are the ultimate games, we can find games for cell phones, and researchers are trying find ways to make games commercially productive. As Julian Dibbell notes, "Ludic energy" is already making people real money. Once the best of all of these technological communication tools find ways to successfully integrate their best-practices – figuring out ways to get gamers to efficiently add to the productivity of big business, and getting big business to be nimble enough to adapt to new digitally-facilitated cultures – how business happens will rapidly and radically change.

Researchers like Byron Reeves have done far-reaching work into how businesses can leverage the computational power of thousands of people interacting within a single online space. We'll soon see that his concept was first introduced by a British anthropologist in the late nineteenth century and is now referred to as "crowdsourcing": something around which venture capitalists are swarming Silicon Valley for opportunities, even in 2008 when financial forecasts speak more of foreclosures and dire times ahead. Few who work in Web 2.0 see the near-term or long-term future as dire; it is full of promise. But what is Web 2.0?

The first iteration of the Web (Web 1.0) is what most people in 2008 are familiar with: websites that advertise or broadcast messages – the places you go to find out something – the ultimate Encyclopedia that you can search on your own terms. Web 2.0, however, began introducing a new aspect to the Web Surfing experience: interactivity.

Interactivity began loosely with simple things like Web Logs (blogs), where the blog-owners post writings or pictures, under which readers of the blogs could post a response. Instant Messaging was one marker of crossing the threshold of Web 2.0, where a computer user can have a small window open on the computer, in which he or she can type messages directed to a single, or multiple members also participating in Instant Messaging. When engaged in this messaging, many windows will tell the recipient(s) when someone is typing, after which, when the typer presses the enter key, what the person typed instantly appears in the window of the person, or people to whom the message was directed. On and on the cycle continues back and forth, creating a message thread...a temporary history of the conversation.

Instant Messaging was a huge hit with digital natives, a "no duh" moment of why shouldn't technology do this? Though Instant Messaging (IM) is better known as the realm of teenage digital natives, the benefits of IM were not lost on large corporations. IBM very quickly adopted the use of IM in its culture, and in a few short years IM add-ons have even leaked into the back offices of financial service institutions like banks, and insurance companies, organizations that provide advanced technology to their customers, but are often very slow in adopting new technologies in their operations because of the cost of giving the technology to everyone in a large organization, and their Information Technology (IT) departments' reluctance to learn and service new systems.

Even though media likes to portray a great cultural divide between teenage digital natives and large corporations, many similarities exist. With large corporations trend setting individuals still work very hard to make a difference, and to be recognized. Early adopters of technology in corporations – those who see the benefits of new inventions – find crafty ways to circumvent corporate regulations in order to use new technology. And just as the implementation of the Univac computers in the late 1950s (THOCP, 2007) was done to speed large and / or complex computational processes, new early adopters of technology today use the technology to speed up their work, to create an advantage for themselves, to make their jobs easier. And when early adopters have greater success at their jobs because of new technology, news of their success spreads faster than a rumor on a middle school campus, creating demand, and encouraging IT departments to provide new technology to others in the corporation.

At this time in our lives, however, technology has always done, and still does nothing more than mirror, or reproduce that which we already do in our society, whether socially or vocationally. What is most exciting is the level of human interaction this is facilitating. Who would have ever thought 15 years ago that thousands of people would be able to simultaneously play games together effectively, which takes us back to some of the crowdsourcing research Byron Reeves is conducting.

The *Wisdom of Crowds* perspective – coined from a recent publication of the same name by *New Yorker* columnist James Surowiecki – stems from the late nineteenth century observations of an English anthropologist named Franic Galton. He was the person who observed and documented the anecdote about how while an individual at a county fair would have a low probability guessing the weight of an ox at the fair, the aggregate of the guesses of all (or most) of the fair's attendees would lead to a guess only a few pounds away from the actual answer.

When social sciences attempt to empiricize such issues, of course there will be holes in the theories, and of course the theories will not cover all possibilities. The concept, however, is being tested in very specific scenarios, with very specific intended results. The concept, when applied consciously, has the potential to be extremely valuable to many people, and is being very seriously investigated by extremely smart people at places ranging from the MIT Media Lab to Stanford University. This is "crowdsourcing."

Allow me to give you the same anecdote that Byron provided at the Virtual Goods Summit at Stanford University in 2007. Remember back to Chapter 2 where I talked about Massively Multiplayer Online Games (MMOGs)? Imagine an MMOG where the base skill required in the game is recognizing pattern anomalies. Now imagine your rank in this game is based on your increasing ability in recognizing pattern anomalies: a one-of-these-things-is-not-like-the-other kind of game. Now imagine that as you increase your ranks and levels in the game, periodically actual x-ray, CT scan, and MRI scan images are introduced as images in which the player is supposed to recognize pattern anomalies. Now imagine that after reaching the highest levels in the game, the most skilled pattern anomaly detectors are the ones actually determining what the pattern anomalies are in actual x-ray, CT scan, and MRI scan images.

The mind-bending results of this "game" are that the results of these skilled gamers, when bundled with crowdsourcing, would technically be at least as accurate as those of trained radiologists. So, theoretically, the future of medicine could see game players paying to make accurate readings of x-ray, CT scan, and MRI scan images. The concept is already playing out on the Web in the crowdsourcing of what is good art and music, and in another well known place: Wikipedia.

Yes, Wikipedia is easy to poke holes in, with the elements of fraud, and difficulty many have in the veracity of much of its content. Academicians like Coye Cheshire – people who take time to really study the veracity of online information – will point out that Wikipedia can already claim one section that is as complete, and accurate as any other Encyclopedia. We must remember that Wikipedia has only been around since 2001, receives contributions solely from volunteers, and is monitored by its peers. With such a loose structure even academicians who study it will agree that there is a great degree of accurate information on something created through crowdsourcing.

While crowdsourcing is frightening on one level, it seems more palatable than having everything sent to Mumbai for analysis, or creation, further reducing the number of jobs here in the US.

Though the Internet changed so much so quickly, **BBC News Online** illustrated that outsourcing – the sending of service jobs outside of the country where the services are needed – would never have grown as an industry without the Internet. (Schifferes, 2006) From computer code to reading radiological films, locations on the globe from Bangalore to the Dominican Republic were able to compete for the opportunity to perform these services required in the United States and in other Western nations.

Media decried the scourge of outsourcing for many years as it tore countless service jobs from the United States economy. After accepting that large American corporations could not compete globally, or even on US soil, if they did not reduce costs by outsourcing these service jobs, the news regarding how large corporations were ruining our economy began to dwindle. News we don't often see involves the countless jobs that were created by the Internet. Before the proliferation of computers in business no one could have ever imagined a company needing to fill roles like "network engineer," "database administrator," "help desk technician" or executive level jobs like "Chief Information Officer," or "VP or Information Services." Some jobs became possible, and entire industries were reinvented because of something dubbed the "Long Tail" made available by the Web.

The author Chris Anderson coined the term "Long Tail" in his book *The Long Tail* that describes how the Internet provides the opportunity to keep a single, unique item for sale, and to make that process profitable. Before the Internet the only place a consumer could find unique items, like out of print books and records, was in specialty book stores or records stores: ones catering to the used market. Access to these items was geographically, and logistically restricted.

The Internet, however, has expanded the ability of large business to offer larger breadth of inventories, and has allowed small specialty retailers to market to much larger audiences. A woman in Flagstaff, Arizona makes her living selling tumbleweeds. Another in Portland, Oregon makes her living selling "Portland candies." A man in Detroit, Michigan runs a record label. All of these people have the opportunity to sell what they do because of the Web.

But the Web is not the magic elixir; just because something is on the Web does not mean that hundreds of millions of people are going to immediately see it. The same issues of exposure exist for artists, as they do for business big and small. A business still behaves like a business, even on the Internet. There must be demand for the products, and the products must continue to be compelling after the initial exposure, or fulfill a specific niche that no one else is serving, which will always be a concern for businesses, whether selling on the Internet or geographically locally. Beyond having good products and services that compel people to either speak fondly of them, or seek them out to purchase again, businesses need to create communication channels that expand the scope of people who know that the business and its products or services exist: Marketing 101.

As Christian Renaud expanded on this topic,

> *People [need] to actually start thinking in a Supra-national type mindset...If you look at all of the myriad of sites on the Internet, that sell boutique chocolate that is made in Portland, ... these guys do [a] booming business, ... chocosphere.com. These guys do a booming business in Portland Chocolate. ... It allows*

these folks access to the global marketplace. So, Neiman Ranch. You've got these little pork farmers in Iowa, that now all of a sudden have [the same] access[to world markets] and they're being sold all over the world. It gives them that vehicle to do so, and think of all of the other gems we're going to uncover in the next twenty-five years. The guy who makes 'THE noodles' in... Indonesia, or Beijing, or wherever, he's [going to] have the global marketplace available to him, and it's going to take us forever to uncover all of these gems. Sort of like the ABE books dot com – the Association of Independent Booksellers – of the world at that point. We're going to digitize, we're going to get access to all of this.

Who would have ever thought about these products – tumbleweeds or Portland Chocolates – would make successful business ventures? But remember, the people selling them are also savvy marketers who have found ways to leverage the reach of the Web.

Computer Mediated Communication has created communication avenues never before imagined. Talk exists of advertising potentially finding its way into cell phone text messages. The digital means of distributing messages is not limitless, but far greater than the days before the Internet. The smarter that smartphone becomes, the more that avenues will open for advertisers' access to consumers. On a smartphone, or web-enabled cell phone, advertisers can already reach consumers when they are nowhere near a magazine, or billboard, or newspaper, or radio, or television. A current television advertisement surreptitiously points to the depth to which advertisers, and content creators will soon be able to reach consumers.

The advertisement's title is "Visual Networking," which, as any academic involved with the social sciences will tell you, directly affects us humans as we are visually oriented creatures. Our visual acumen, while not as acute as other creatures in the animal kingdom, has kept us alive for millennia for our ability to differentiate friend from foe, and edible materials from non-edible: safety from danger. In "Visual Networking" a street scene starts with a man receiving a visual how-to over his cell phone to fix his scooter. The scene then pans to follow a woman walking her dog. When she and her dog near a visual display kiosk, which is one wall for a bus-stop shelter, some form of sensor recognizes that a person and dog are present and display a live-action short-form video advertisement for dog-food.

The advertisement continues on to illustrate other near-term future possibilities with video that allow even greater levels of Computer Mediated Communication: those that allow friends to recognize and communicate with each other from different street corners (which could be blocks, or miles apart), and those that help us navigate from one point to another. Really everything in the advertisement is already here.

Skype video communication allows real-time video conferencing from your computer, which is currently free to users. Facial recognition and other systems allow computers to "see" the world around them, and respond to that world based on what it sees. Videos long and short can already be viewed on cell phones. Small computers tied to Global Positioning Systems (GPS) already direct us from point to point in our cars: either factory installed, or portable versions like the Garmin Nüvi.

The advertisement is not science fiction; it's taking already existing technology and weaving it into our daily lives, making it more ubiquitous, and allowing it to touch us at every street corner.

Other current technologies allow us to touch each other virtually today. When compiling the interviews for **IMHO**, I learned of a conference that I desperately wanted to attend. The problem was that my wife and I had already scheduled a family vacation to Hawaii with our two boys during the exact same weekend on which the conference was occurring. Being about virtual worlds and the communication advancement virtual worlds were making possible, the conference was also going to be held in Second Life. With this news I thought, though I would miss meeting face-to-face a few of the people with whom I'd been speaking, I'd still be able to attend this conference in Second Life, and see exactly how effective such attendance would feel. So along with flip-flops, and sunscreen, I packed my laptop loaded with my Second Life interface and screen capture software.

Wondering about how this was really going to be useful I thought about the IBM campus on Second Life. In the center of the virtual campus is a large amphitheater capable of seating over 280 avatars. I thought, "they must get some serious use out of that…they took the time to create it." So I blindly continued forth on my quest of attending a virtual conference. Interestingly, folks at IBM admit that they have yet to begin tracking hard numbers regarding how much money they save using virtual worlds.

Karen Keeter, a Marketing Executive at IBM noted "There are some 'back of the envelope' type calculations you can use to figure out the cost for bringing people to a live event - determine the average cost per person per day to bring someone to a live meeting, multiple that by the number of people and the number of days, add in a productivity hit (time away from office) for extra credit, and you have an estimated cost for a live event."

For a one day event in their 300 seat amphitheatre that could easily be a cost savings of:

Airfare	$400
Accommodations	$150
Food	$100
Productivity Hit	$ 300 (based on $75,000 annual salary and 240 potential days of work)
Savings per Employee	$950
x 300	$285,000 savings per day

Saturday after we arrived at our hotel the conference had already been going on for a day. With the time change between Hawaii and Palo Alto, CA, I had to wake a little early and carry my laptop downstairs to an area of the hotel with free Wi-Fi access. A couple of Alt+Tabs between screens on the laptop, a couple of clicks, and my avatar teleported to the conference site. I woke a little too late on this morning so by the time I joined the conference it was already in session, so I had my avatar find its way to an open seat in the virtual amphitheater, and sat it down to take in the conversations around me as well as the streaming video of the conference's speakers in Palo Alto, CA.

The experience was not perfect. My laptop is older, so my graphics card was struggling to display everything on the screen. I went into the Second Life settings to tailor my experience for this scenario. If not explicitly told to not do so, Second Life displays little bubbles above each avatar which lists the name of the avatar, and any affiliations the avatar might have with groups within Second Life. I stopped displaying those. That helped a lot. I had to tweak another setting for how the video displayed, and then the streaming video came through well...not perfectly, but well enough to tell who was talking, and see how they were interacting. I saw enough of the nuances that it was obvious what was going on.

But how did I know to do this? It was not published, and I'm not such a computer genius that I intuitively knew to search for such remedies. I looked around the room, and watched the left corner of my screen that kept a running history of all of the things all of the avatars in the room were "saying," or at least typing. Even though I was late, others were late too, and upon arriving people would ask about "lag" or "refresh rates" or "contrast." Some virtual attendees were asking about why a presenter was say something, and then we would see the corresponding visual cue to which they were speaking a second or two later: the lag. Others would ask why the video on the virtual amphitheater's JumboTron-type video screen would stutter or appear as a gray box: the refresh rate. Others would ask about why areas of the amphitheater were practically un-viewable, lost in either dark shadows, or awash in too much light: contrast. Quickly other attendees would type suggestions. The community there wanted everyone there to have a positive experience. Everyone wanted to help each other out. People who had answers wanted to publish – even if only fleetingly – their answers, subtly broadcasting their command of the medium, and inferring their expertise.

While not the same as attending a conference where you have the flexibility to see and hear someone, then decide at the last second to go up and introduce yourself to that person and have an impromptu conversation in a hallway, I still had the opportunity to listen to what was presented at the conference, and to interact with the participants in our little corner of the conference. The participation was funny, too. It was like a middle-school free-for-all at times. People threw out comments almost stream-of-consciousness while listening to the streaming video, and from those comments tangents would sometime spawn that related to the initial comment but like comments that strike a social nerve, take off on their own. The difference, however, was that occasionally the speakers on the streaming video would react to the comments; they were able to see, read, and respond to the comments being made.

This interactive voyeurism steps into a discussion arising in Computer Mediated Communication spaces: the difference between privacy and secrecy. Some of the avatars were obviously broadcasting all of their comments publicly, with the intention of others reading their comments. There's no way the users behind the avatars could have been typing that quickly in order to write what they were, and maintain private conversations. In Second Life you can engage a person in private Instant Messaging only viewable by participants you target, while general chatting is observable – consumable – by anyone within "earshot," or a predefined virtual distance. At this conference other avatars were sitting unnervingly still. They were either sitting quietly and consuming everything going on around them, or they were engaged in private IMs. So what compels someone to have a private conversation? In this new realm of digital interaction, what is private, and what is secret? It's the issue of the changing "publics."

Interestingly enough, a study in a small village in the UK illustrates that when provided access to information about their community – in essence restricting privacy in public spaces – citizens not only ask for, but clamor for more. Jamais Cascio noted:

> *...the first reaction people have when you talk about all of these cameras all over the place is 'what about my privacy?' ...People seem to be more comfortable with unaccountable authorities peeking in their windows than everybody being able to watch everybody else. Interesting experiment, though: in the UK there was a small town that decided to open up their cameras and make it so that everyone could view all of the data coming from, or all of images coming from the public cameras. And rather than it becoming a situation where people got freaked out about their privacy, they*

wanted more. They wanted to see more. They wanted to have even greater access to the video streams. And whether that's a peculiar condition of that particular culture – small town in England – or that it's something indicative of where larger communities would go is currently an open question, but a very provocative one.

The boundary between private and public spaces is being redefined. What happened in the small town in Britain has been surreptitiously happening here in the US for years. Privately and publicly owned surveillance cameras have been installed since the inception of closed-circuit television. Only now we are able to create an easily accessible network of these cameras, which while problematic to some, provides great opportunities in the realm of protecting citizens again terrorists.

New York City is already on the path of making this web of surveillance more than dumb cameras that can show us what happened after a catastrophe. The May 2008 issue of **Wired Magazine** has an article that describes the integration of private and public resources the New York City Police Department is deploying in order to thwart terrorism.

The article titled **The Shield**, by Noah Shachtman, describes how London is already one of the most surveilled cities on the planet, yet that surveillance was unable to stop the suicide bombers in 2005. (Shachtman, 2008) What the article goes onto detail is how The City Planners in New York are integrating building inspection resources, private resources, and public resources, into a network layered with newly developed video monitoring software. The efforts of all of these previously disparate entities are part of a plan to create a city-wide video surveillance network by 2015. The influence is so wide-spread that blueprints for new construction submitted to the City are reviewed to insure that the designs take into account security concerns. In one high-profile case, the Freedom Tower was moved further east to address security concerns.

Most critical to the plans and implementation is the incorporation of new and emerging technologies: facial recognition programs, and systems that recognize patterns. Instead of creating an enormous network of dumb camera – systems that can only record, or allow trained personnel to monitor a limited number of cameras – images from the network will feed into these new and emerging technologies, and, hopefully – the article emphasizes hopefully – be able to recognize terrorists while they are still planning their missions, not just record the execution of their missions. Even more interesting is the cooperation of private corporations that the City has been able to garner. Of the 3000 cameras that will be in the network in Lower Manhattan – just in Lower Manhattan – two thirds of them will be privately owned. (Shachtman, 2008)

Ethically, speaking to part of the new cultural discourse that has arisen from events like September 11th, and the opportunities provided by Computer Mediated Communications, this is part of that conflictive space where new norms are developing.

In the Jamais Cascio example of the small town in Britain not only welcoming the access to public information, but clamoring for more, is the question more the blurring line between private and public?

As danah boyd says, "what are the new publics?"

When a Second Life avatar is saying something to another avatar that is not part of a private IM, do those words become public property? After all, what the avatar "says" is capturable, and repeatable, infinitely copyable through numerous digital tools. The digital tools continue to reach deeper and deeper into the spaces we previously considered private, which creates both ethical dilemmas as well as commercial opportunities.

As the Web continues to develop, we see new frontiers unfolding. The first websites and Web 2.0 created a digital platform for companies to display products. Innovative companies exploited technology to showcase products and services through what was termed multi-media (the integration of text, still images, audio and video). Platforms like Second Life and There.com illustrate a few of the possible digital frontiers for business. But the frontiers do not have to be as fanciful as 3D virtual spaces.

In the television advertisement scenario described earlier, we see interactive billboards. Think about the blurring line between public and private in this scenario, which is not limited to corporate thinking. The Tom Cruise film *Minority Report* illustrated the use of interactive billboard technology. Imagine walking down the street and having the advertising medium in front of you at first recognize that you are either a man or a woman, and then modify the advertising it is displaying based on that. As that technology progresses the advertising medium will posses other pattern recognition tools that allow it to see what kind of clothes you are wearing, your height, your gait, your affectations, and – while based solely on stereotypes – possibly better target the advertising to you.

In today's terrestrial landscape we all expect to be bombarded by advertising, but it is static, or at least broadcasting one message to the entire community. In this not-so-distant-future scenario advertising knows who you are. In the most futuristic of scenarios like in *Minority Report*, the advertising medium will actually know who you are, not just make an educated guess at your likes or dislikes. To make all of this technology work will take the work of a lot of corporations making strides on their own and working together. Some will make a lot of money from the attempts, and others will not. This is occurring today in the virtual landscape.

Many companies exist that focus on creating a presence in a virtual world for other companies. Electric Sheep and Millions of Us are two of the consulting companies at the forefront of the virtual world revolution, creating new environments in Second Life and other virtual worlds for Fortune 500 companies and others. But the initiatives the companies develop are not always specific to virtual worlds; it's about communication.

Rueben Steiger said,

> *My bet… is not whether people want to be in this [virtual space], or that [virtual space], or whether this one will beat this one out, but it's that people want to connect with other people around shared interests. And second to that, that the more immersive the format in which they can do that, the better in many cases.*

What is attractive about the virtual worlds, however, is how easily a company's story can translate. As Sibley Verbeck commented,

> *…it's very easy to tell stories about what's going on in virtual worlds, and therefore it makes a better press story, and it helps companies – whether it's*

entertainment companies, or companies thinking about using this technology – understand it more easily. Because even though under the hood it's very complicated, and it's still very unclear what directions technically this [virtual world] industry is going to go, it's difficult to make technical choices for projects – never the less, on the surface it's a very human technology. It's even easier to understand than other new communication media that come around. I think even easier to understand than the Web when it was starting.

The headlines – often appearing on the business pages of newspapers – range from curious intrigue to accusation. If you are not paying attention, it's easy to not notice how much press virtual worlds, and our digital existence receives.

Monday, June 11, 2007. **The Boston Globe** Business & Innovation Section. **New social website tempts the inquisitive**, by Carolyn Y. Johnson. (Johnson, 2007) "The online social-networking world is crammed with websites where college friends, nurses, moms, and even cat lovers can mix and mingle. [In this one] it's not who you know, but what you know."

Monday, July 23, 2007. **The New York Times** Technology Section. **Cute Friends to Collect, and Plug In to the Internet**, by Daniel E. Slotnik. (Slotnik, 2007) "The Webkinz stuffed animals are not only cure to line up on a bed, but they also unlock a succession of fun places in cyberspace…"

Thursday, August 9, 2007. **The New York Times** House & Home Section. **A House That's Just Unreal**, by Seth Kugel. (Kugel, 2007) "From the roof deck of Sherman Och's Mexican-style villa atop a breezy bluff, the entire island of Jalisco, population 20, spreads out below. … Jalisco is a sim (for simulator), a plot of land in Second Life."

Tuesday, November 13, 2007. **The Contra Costa Times**. **Parents say no to computer games**, by Alan Fram and Trevor Tompson (Associated Press). (Fram & Tompson, 2007) "Those surveyed in poll seldom play with their children."

Sunday, January 6, 2008. *The Sunday Times*. *Cyber schmooze is not just for geeks*, by Mark Abramson. (Abramson, 2008) "Psychologist says going online to meet people is more socially acceptable."

Tuesday, January 8, 2008. *The Contra Costa Times* Time Out section. *Virtual affairs of the heart: Cyber infidelity can be as damaging as the real deal*, by Jessica Yadegaran. (Yadegaran, 2008) "In real life he is a successful 35-year-old business owner and husband-to-be. But on Second Life, the virtual fantasy world with 11 million 'residents,' his avatar, Lugh Dragonash, is a cyborg, or human machine, which can make it difficult to meet women he says."

Monday, April 14, 2008. *The Contra Costa Times*. *When teens go wild, we blame the Internet*, by Susan Young. (Young, 2008) "We used to blame the TV for all the moral sins of the nation. Now we can put the blame where it really belongs: on the Internet."

Thursday, May 8, 2008. *The New York Times* Circuits Section. *When Web time Is Playtime*, by Warren Buckleitner. (Buckleitner, 2008) "If you've been noticing an increase in the number of smudgy fingerprints on your computer screen, it may be because your young children are spending more time online."

Even with the periodic negative press regarding teenage (and even adult) behavior on the Internet, most people involved with the medium recognize the shifting norms: the acceptance of personal and financial transaction on the Web. This is not lost on big business. CNN, Scion, Coca Cola, and other Fortune 500 companies have all started to, or already have created a presence in virtual worlds like SecondLife. IBM and Cisco have entire campuses created in Second Life where they regularly hold corporate meetings.

IBM has the huge amphitheater where avatars can come, sit, watch, and participate in conferences. Cisco has private conference rooms to which you need an invitation to visit. Christian Renaud invited me to one of these conference rooms that was appointed with groovy furniture and wonderful vistas. Yes they were all digital make-believe, but they were still compelling, and worked to create a more comfortable environment in which to speak with Christian.

As I had mentioned in Chapter 1 regarding what Byron Reeves said,

*...you may be able to think your way around that
similar response [– the neurological responses we have
to a virtual experience being the same as a real world
experience –] if you spend a lot of time. If you are in a
virtual world, someone comes up to you and says
something bad, or does something you don't like, you
may be able to think 'well gee, this is just media, these
are just pixels on a screen, it might be controlled by a
computer rather than another person, I shouldn't take
offense, I shouldn't be excited. But that's really a very
difficult response, and probably something that's not, by
any means, the default.*

And what's so wrong with having the suspension of disbelief
when immersed in one of these worlds? It is the same as the time we
spend in the movie theatre where the majority of scenes are digitally
enhanced. Even before digital enhancement of films we suspended our
disbelief to watch King Kong climb to the top of the Empire State
Building, or watch the USS Enterprise boldly go where no man had
gone before. We look back on some, not all, of the model-based movie
making magic and chuckle at how infantile it now appears, but nothing
like it had ever been done before. Fifteen, twenty years from now we'll
be looking back at how poorly we rendered human facial expressions,
and wonder how anyone believed any of the digital effects we watched
in movies. But we are doing a great job today for more fantastical
creatures – ones that require a suspension of disbelief to merely exist in
the story. How else would Hogwart's Castle exist a mere train-ride
from London?

And here lies the greatest boom to business. Where reading
radiological films has a component to it that is enticing to people who
are already gamers, the digital frontier does something more appealing
to those of us not as thrilled by puzzles and chases. After meeting with
Christian in the private office to which I teleported, I had my avatar
teleport back to the general Cisco campus that's open to the public. I
wandered around and found a lovely space fashioned to look like a
Japanese teahouse. I had my avatar sit in the space adorned with a koi
pond and candles, and I could feel myself relax. I turned up the audio
on my computer and heard that the designers had actually put the

gurgling sounds of a water fountain into the background noise for this space. I did not totally suspend my disbelief, I was too self-aware of what I was viewing, but I did empathize with my avatar, as I would a character in a film, and allowed myself to share in my avatar's experience.

At this point I could have easily gone off to do a "job" in SecondLife, as some people do. There are tasks and activities that users can have their avatars do that will earn the avatar Linden dollars. Some of the jobs are as simple as "camping," having your avatar lie on a couch or beach chair. These tasks don't earn a lot of money, but the simple act of hanging out around friends or strangers, and chatting with them can earn you virtual currency that you can later use to buy your own virtual island, and build your own virtual house, and new tricked-out virtual clothing, or even a virtual car. Remember back to the Chinese gold farmers. The inventiveness, and intensity in which avatars (or is it the users) pursue money-making ventures is solely up to the participants, just like real life.

As mentioned earlier the *CSI* franchise has a space in SecondLife. I met a woman at the gateway to the *CSI* sim (simulation or simulator in Second Life) whose in-world job it was to greet people as they arrived at the *CSI* sim. This was her avatar's job, for which she was compensated: making sure the avatars that (or should it be who) arrived at the *CSI* sim had an easy and fun experience at the sim, facilitating their hopeful return. What's magical about these worlds is that you can immerse yourself in the aesthetic you find most pleasing, or productive, or stimulating, or maddening (if you have the need to feel more anxiety in your life). The Weather Channel has a sim complete with a mountain for snowboarding, a desertscape for mountain biking, and a beach for surfing. Your avatar can do any of these activities in this sim. One of the avatars I met in the sim even admitted to surfing while on conference calls.

Are these sims perfect digital simulations of what they are trying to represent?

No, and they're really not trying to be. As we've seen, the suspension of disbelief is a far more compelling tool for where technology is today. In the near or mid-term future, technology may reach a point where we cannot tell the difference between what is "real" and what is digitally constructed in an augmented reality plane, but for now we are not there. In a conversation with Henrik Bennetson from the Stanford Humanities Lab, we discussed the recent movie *Beowulf*. The movie was neither live-action, nor pure animation (computer or otherwise); it was a hybrid of live-action and computer animation, where motion capture of actual actors was digitally morphed with computer animation, making the characters more disturbing than compelling. The movie crossed that border that allows us to accept the suspension of disbelief.

Within these new digital frontiers, however, the borders are hypothetically, and practically limitless. Think about humans' potential for innovation, and digital natives who look at these new tools not from the perspective of their limitations, but from their possibilities, and you have the threshold of a new World, with a capital "W": a place where a person could easily "Live" his or her entire life, with the exception of eating, or drinking, or sleeping. For the Chinese gold farmers this is already happening today, and for those who have figured out how to do this – those who have consciously pursued this – their work is their play; they have blurred the lines between the two. Think of the new generation of gamers coming along. Seeing the process of gathering currency as simply playing another or new game, their mindset is already prepared for this meshing of what we previously considered two very disparate endeavors.

Chapter 6: New Borders

"The Internet has evolved into an important industry in China, exerting a great influence on people's daily lives and everyday economic activity and helping promote the country's economic development and social progress." – *china.org.cn*, November 28, 2007.

In Chapter 5 Christian Renaud's comments about thinking supra-nationally relate to how many undiscovered products exist in the world – for example "the guy who makes THE noodles in Indonesia, or Beijing" – which is only a small element of how the Asian Pacific rim is going to influence technological innovation, content creation, and content distribution. China is a country that is not only poised to take a leading role in Information Technology management and development, they are confident that they will be the world leader in this field. While other countries in the Asia-Pacific rim may not have the same bravado as China regarding their near-term roles in Computer Mediated Communication, their current rate of development and adoption of technology is far more mature than the United States.

Chi Tau "Robert" Lai was a casual attendee at the Virtual Worlds Conference 2007 in San Jose, CA. He was there to solidify relationships he had been nourishing with a company called MindArk, which owns, operates, and develops a virtual world called Entropia Universe.

Robert is an unassuming character who came to the Conference with modest expectations. John Bates, the resident evangelist for Entropia Universe encouraged Robert to come to the conference because he knew that people with vision at the conference would not only want to hear what Robert had to say, but would seek out Robert. And seek Robert people did. He was a cult celebrity among the technology marketing geeks. After all, this was a trade show where all of the industry leaders gathered to show off what they were doing, and how they were extending the frontiers of this digital universe.

One evening at a cocktail party held by one of the conference sponsors, Robert expanded upon an analogy he repeated many times during the conference to many different micro-audiences. "China," he said, "has been making shirts for the West for years. The shirt that sells here for $20, we make in China for $1. We want the other $19."

Robert sees virtual worlds as the gateway to the other $19, and he is not alone. Of course this vision is held by other evangelists like John Bates, the current corporate evangelist for MindArk, and company heads like Rueben Steiger, the owner and president of Millions of Us who noted that the digital future is not about where people interact, but that they want to interact; there will always be a market for the digital tools to facilitate that interaction, tools that will always be changing.

But listen to Robert for a short period of time and it's easy to drink his Kool-Aid. He doesn't come to the digital universe from the avenue of techno-babble; he talks about how people work, how people think, and how people use technology. After all, regardless of how flashy any technology is, how easy it is to use, how useful the developers think it is to humanity, if no one uses it, if digital natives don't see reason to embrace it, and make it do things the developers never considered, then the technology is as meaningful as a broken brick doorstop.

Robert sees so much promise in the MindArk technology behind Entropia Universe that he, on behalf of the Beijing Cyber Recreation District (CRD), has partnered with MindArk to use their "engine" – what makes Entropia Universe work – for the creation of the virtual universe that will be the Beijing CRD digital universe. So what really is the CRD?

Direct from the CRD website,

> *'China Virtual Economy District' it will play the leading role of virtual economy industry in the world. With a series of trading rules and data exchange standards, virtual enterprises with different resources and advantages can cooperate and help each other raising their abilities in doing business. 'China Virtual Economy District' creates a wide and whole new platform for traditional enterprises, internet (sic) companies and individuals to participate in. This means that the dream of the inter-connection of virtual world with real world will come true, and the new generation of culture originality of CRD is started. Also means that this is a giant step to the promotion of industry infrastructure of Beijing Shopping district. (CRD, 2007)*

After reading this I would start taking Mr. Lai's anecdote of reaching for the other $19 very seriously.

What makes CRD's partnership with MindArk interesting is currency. The business people at MindArk purposefully created a game where the in-world currency is easily convertible into "real world" currency.

Look more closely at Entropia Universe, which is MindArk's virtual world. On May 2, 2006 in the *New York Times*, an article about Entropia Universe by Seth Schiesel spoke about a project MindArk was planning on implementing: an ATM card that allowed users to liquidate in-world currency for real world dollars. So after selling an in-world item – for example, something like a dragon saber, but this is not accurate in Entropia Universe, which deal more in weapons like laser rifles – and receiving in-world currency (PEDs in Entropia Universe), a user could use his or her ATM card and receive that balance in real world dollars.

Entropia Universe is designed to provide a compelling 3D in-world experience, which speaks to the comments from Byron about turning up the volume on a user's experience, and Sibley's comment about virtual worlds being an easy story to tell. Then there's Entropia Universe's interface already set-up to manage transactions with real-world currencies. 2005 was a big year for Entropia.

Later in the same *New York Times* article you read about exactly how mind blowing that 3D experience is, particularly in real world dollars. The article describes the story of one user, a man named Jon Jacobs, who liquidated most of his real world assets, and gambled by buying an in-world space station for $100,000. That's right, 100,000 US dollars, which was equal to 1 million PEDs. By subdividing this property, allowing other users to develop on his "property," and taxing the activities of the users who developed on, and otherwise used this space station – something users were already accustomed to in Entropia Universe – Jon Jacobs was able to make close to $12,000 a month from his investment, and he had not fully developed all of the "real estate" that he had purchased in-world. (Schiesel, 2006)

All of this real estate development was real, except for there not being any dirt, or wood, or metal, or real world tangible materials that a person could touch, or smell. Everything, however, was very easy to see. Like looking at pictures of someone building a house for you miles from where you live, the mind doesn't care that the objects are untouchable; as far as your eyes are concerned, the properties, and buildings are very real. And through this manipulation, and construction, Jon Jacobs was able to make some very real profits.

These were real world, US dollars. Entropia used the efforts of Mr. Jacobs to their advantage, by gaining much exposure through press like the *New York Times* article, but also by legitimizing the economic viability of what before were obscure gold farming operations. Virtual Worlds were now a new frontier for real commerce.

If China's virtual world experience was more of a game than a transactional tool, users would have to add an extra step of using eBay or other bartering mechanism that grew out of the first generation of virtual worlds. The CRD has eliminated that step by choosing the virtual world engine that, on the surface, is a fantastical science fiction environment, but is already built for commerce. Robert Lai saw that the characters and landscape were window dressing. What will be interesting in the near future is to see the window dressing the CRD chooses.

What's important about this is Robert sees what Christian Renaud is saying. Robert sees the promise of Reuben Steiger, of John Bates, and of all of the other industry folks who see a lot of commercial promise on the horizon for the Metaverse. They see that technology is almost secondary to the pursuit, that to be successful means being flexible and using the technology best suited for an immediate need, not trying to wager a bet on what will be the only tool that people use for communication in the future. They see that digital natives will take the technology given to them and use it in ways the developers never intended or thought possible. And Robert is not just blinded by an authoritarian vision of the future coming from a polit bureau. Robert came to the United States in the 80s, graduated from Carnegie Mellon University, stayed on to teach at Carnegie Mellon while beginning down the path of a PhD, then, like so many PhD candidates, saw growing opportunities outside of academia. He had the opportunity to stay in the States, but returned to China to help this country become what he sees as possible, just as many Westerners have moved to China during this time of enormous economic expansion in the country.

Listening to Robert speak was compelling, but so was observing how others dealt with him, and how little many in the West know about Chinese culture, and how the Chinese view themselves in the context of global business. Many Westerners at this Conference appeared to look towards Robert as a novelty, looking for the place where they could sell their wares into his country of limitless possibility. Billions of potential customers! How intoxicating. Robert, however, would drop kernels of information through his conversations, illustrating how the Chinese government was using the Beijing CRD as a form of incubator – a place to create a powerful, stable computing environment in order to further their plan of rapid economic expansion through digital frontiers.

China is neither capitalist, nor democratic. One marketing executive naïvely asked "How do you deal with, and integrate multiple computing platforms?" Robert chuckled. "We don't integrate, we regulate!" And following Robert's thread through the remainder of the conversation you learn that what plays out in the CRD is what will be the regulated blueprint for government controlled infrastructure rolled out through the remainder of the country.

This, of course, raises many larger questions, not regarding the power of technology, but of the power of information. As Chinese nationals receive access to more and more information, what will stop the increase in black-market technologies within China that allow for access to censored information? Will China's great information technology plan be its ultimate undoing? Will capitalism and democracy swell from the potential access to what Christian Renaud called "the corpus of human knowledge?"

After all, if we follow the path of digital natives, isn't the path of China from authoritarian regime to democracy practically inevitable? The most exciting (or frightening) possibility of this question is the economic force that China will unleash if they maintain the path they are currently on.

As Jamais Cascio comments,

> *China seems a pretty powerful counter-argument to the notion that access to technology and information leads to democracy, and to the Friedman, **World is Flat** premise that development is impossible without democracy.*

Jamais continues,

> *the notion that true development can only happen in a democracy...Now, depending upon how you phrase [development], true development can only happen when members of society feel that they have some power to make changes to their life...then that's something that has been happening in China, even if the*

traditional, our traditional concept of democracy hasn't really played out very well, or very thoroughly. The growing capacity of individuals in rural villages, in urban settings to be able to make choices about their life has been a real engine for growth and social evolution in China. Whether that leads to more democracy down the road is actually quite uncertain, because you have a lot of people there, a lot of people there who are very happy with what kind of changes they've seen, and very happy with the fact that it's come from an essentially still authoritarian government. Similarly you saw real advances in South Korea under a fairly authoritarian regime. Now, it did transition to a real democracy, and that's terrific, but it doesn't necessarily have to go that way.

The Robert Lai anecdote touches on practically every aspect of how technology can facilitate the dreams of a society. How emerging economies will manage creative content is a large part of this. In a mature economy like in the United Sates, where the Internet is a communication medium and a conduit for commerce, content is king. The intention behind this cliché is that without something to read, see, hear, or otherwise consume, something other than a list sheet detailing what is for sale on a website, without the content, there is no reason to visit a website. Without content, there is no reason to turn on a phone. Without content there is no reason to turn on a TV. Of course some of these statements are silly and self-evident, but Computer Mediated Communication media are different from other terrestrial activities.

Creative content produced in a medium like a physical book is something that you can also touch, smell, and hear. One could argue that you can taste it too, but that's a fetish I'd rather not explore. The tactile nature of a book printed on paper, deepens the experience through the sound of turning pages, the smell of paper, and the feel of a book, from its individual pages to the cover.

While the content of that book can also be distributed electronically through channels like the Internet, or new eBook readers from companies like Sony and Amazon, the experience of consuming that book is different than when it exists on the printed page. The experience is not necessarily better or worse, it is simply different. The difference in experience starts with finding the book. In an electronic format we can make pinpointed searches to find that book, or its contents, through search engines with ever increasing capabilities. From Amazon.com to Google, finding digital records of the printed word is becoming easier and easier while the outlets for finding that content are becoming more and more numerous.

So why would someone choose one method over another for finding the printed word. Maybe one indexes and provides access to more books, papers, and magazines than another source. Maybe a different source has better content to allow consumers to discover works they never would have before considered: content like book reviews, and chapter previews. Perhaps yet another source is the best resource for books and articles about a specific topic. In either scenario, the end result of finding a book to either download, or purchase for shipment has other content involved in order to bring consumers to that Computer Mediated Communication medium.

In all cases this is a conversation over consumers and their consumption of media. The word "consumers" does not necessarily equate to a person actively or passively looking to be the participant in the trading of currency for goods or services. Consuming from this perspective is the taking in of media, visually or aurally.

While consuming a ham sandwich has very palpable results, the consumption of media is harder to measure, and Computer Mediated Communication is making this even harder. The number of potential access points for consuming media through computer mediated outlets is not only growing rapidly, but infiltrating almost every corner of our culture. Rarely does someone have to pursue media; a consumer usually must make a conscience decision to avoid it, or to turn it off. Many countries in the Asian Pacific rim are adding to this growing access to media, as well as to the quantity of media content, much to the disappointment of content creators in the West.

Hollywood has long exported American culture to other nations, so we in the West have become very accustomed to viewing the Americanization of other cultures. Because we know what McDonald's is we find it easy to understand how a McDonald's might appear in Paris or Tokyo. The same goes for American words and phrases that creep into other cultural lexicons; things viewed as quintessentially

American: anything from "Thank you very much" a-la Elvis to a Terminator's "I'll be back." Computer mediated communication, however, facilitates the movement of culture in both directions. While Western culture has long seeped into regions ranging from East Asia to Western Europe, we are slowly seeing the insidious creep of cultural habits from East to West.

Chapter 2 contained a brief illustration of the nationalistic cultural norms surrounding Internet gaming in Korea. This sense of national pride and unity is not unique to Korea; it is so elemental to many Asian cultures that interactions between people from different Asian nations can take on xenophobic tones. With the strong sense of nationalistic belonging, however, comes a reduced focus on the self. Note the recent events with the Olympic torch relays leading up to the 2008 Beijing games. Protest after protest broke out around the globe, following the path of the torch, and vilifying Chinese policies regarding Tibet. In China, however, blog postings exploded with pro-Chinese rhetoric supporting Chinese nationals, and Chinese policies, which is not surprising given Jamais' comments regarding culture and geography.

> *Majority cultures tend to strengthen the geographic aspect [of their culture], so you have...Xenophobia erupt. You have the examples of the Chinese bloggers with the reactions to the Olympic torch chaos, where the reactions from the Chinese bloggers weren't just 'these guys are screwed up' [referring to the critics of China], but how dare they insult the great Han race? They're trying to humiliate China. It's very much a nationalist, jingoist, borderline racist response that was strengthened by that kind of Internet communication capacity.*

With the era of modernism in the West brought on by the birth of the novel as a written form, and other epistemological changes, the late eighteenth century and beyond saw an ever increasing importance on the individual, but many Asian cultures did not see such a change. Culturally many East-Asian cultures see its people in terms of a collective. Accordingly, intellectual property in the East is more often thought to be the property of the collective than the domain of the individual who produced the content.

The current powerhouses owning creative content in America are now faced with a previously unseen dilemma. The Eastern cultural view of creative content, where content is the property of the masses, predominates the Internet, not the Western view of ownership held by the individual. Content being king, therefore becomes a difficult business proposition as consumers of media expect free content. The foundation of the Internet is that of a network of computers that can easily share data. What we know as the Internet today is more of a server and client relationship, one where one computer (the server) contains a store of information that is served to the Web. So when you type in an Internet address your computer is really finding and seeing the files that reside on the computers that are related to that Internet address.

The initial attempts at creating the Internet, however, did not have this typical client / server relationship; it was based on a peer-to-peer structure. One computer was linked to another computer, and the two computers were able to share information. When this is done in your home or at your business the relationship is often referred to as a network. That initial concept of peer-to-peer information exchange is always lurking among creative users of the Internet. One well known manifestation of this was called Napster. When Napster first started operating on the Internet its popularity exploded. Anonymous users all over the world were able to instantaneously share their music, pictures, videos, and anything else that could be digitized and stored on a computer. Without well-known methods of marketing or promotion Napster became a tool of rampant piracy. But it's not that all of the people using Napster were thieves; it's that on the Internet, the collective zeitgeist sees content as free and interchangeable: the domain of the masses.

Content creators in the West struggled for years with the ethos of the vox populi. People share. Community is often about sharing. When a person whose taste a community respects finds a new song, or book, or film, the community wants to share in the experience. Families share books. Teenagers in the 70s and 80s made and shared cassette tapes of their favorite music. Films, up until now, have been problematic for sharing.

When teenagers – mostly college students with access to fast

Internet connections – simply took the age-old pastime of sharing into the digital age, music publishers took flagrant violators of copyright laws – the largest downloaders of "free" music from Napster – to court. The laws had always existed, but the process of copying was so arduous that music publishers actually found the sales of albums rose when there was a small degree of this piracy. The cassette tape medium, after all, degraded much more quickly than digital copies of music do, so if a friend received a copy of a record album, and really enjoyed the music, that friend would most likely go to the local record store to buy his or her own copy. In the digital age real consequences became associated with infringing copyright laws. Napster was shut down and music publishers clamped increasingly restrictive tools on their digital media to inhibit piracy; this is what I earlier defined as an example of Digital Rights Management. While music publishers believed this restricted access to music and the prosecution of the illegal downloading of music would reinvigorate their sagging sales and traffic to their websites, content consumers found other ways to pirate music, and continued to shy away from purchasing music through orthodox channels. Along came iTunes, which allowed consumers to buy music one song at a time, with DRM that pleased the music publishers, and the music industry changed forever.

With strengthened copyright laws and new definitions for protecting copyrights in this new medium, newspapers, online magazines, and other print sources began restricting access to their sites. *The Wall Street Journal*, *The New York Times*, even *Salon* – an online magazine that did not exist before the digital age – all started to charge subscription fees for access to their content. (Schonfeld, 2008) Consumers sent a very loud message to these content providers and instead frequented sources that were free. While these sources of content no longer charge for their content, many now ask consumers to register on their site in order to access the content; this registration serves a few purposes for the content creators, one of which is tracking the active accessers of the site, which helps the providers gauge what to charge advertisers on their sites, one of the greatest methods by which these content providers make money.

How to monetarily compensate the creators of creative content begins to be a different and still-changing question. Before the growth of content for the masses, a limited number of gatekeepers existed in the creative content industries. Music, movie, and publishing executives created processes that limited what, where, and how creative content made it to the masses, giving the masses the privilege to purchase individual copies of what the executives had deemed as tasteful.

From laptops to smartphones how this content is distributed is rapidly changing along with how it is getting created, by whom it is being created, and how it is being labeled as tasteful. The gatekeepers no longer have the control they once did. As Chapter 4 discussed, the power of popularity is moving farther and farther into the hands of those consuming the content.

And this means great opportunity for the Asia-Pacific Rim. Like the company selling Portland Chocolates, Leinad Zeraus, self-published a science-fiction novel called *Daemon*, the sales of which gained traction through word of mouth and the Web. The Internet will allow consumers in the East and West to determine what the best value is, what is good, and to rapidly distribute that information to others. Mainstream literature publishers have been taking advantage of this development for a number of years.

Technology allowed for the creation of on-demand printers a number of years ago, like iUniverse, Lulu.com, or Amazon's Book Surge, allowing unpublished authors with no agent, and no access to the orthodox distribution channels, a way to publish their books without bringing their manuscripts down to Kinko's, and binding their creations with heavy card stock and tape bindings. Instead, for a nominal fee, an unrepresented author could have one copy of their book(s) printed at a time. These were professionally produced paperbacks that looked no different than books coming from larger literary publishers. With a little effort, heavy networking, and clever leveraging of Internet marketing tools, many of which were free, an unknown author could build a groundswell of enthusiasm behind his or her work(s). Anecdotal evidence shows that during that time, and continuing today, an unknown author who was able to sell 600 to 1000 units of a book in an industry-judged short amount of time, without the support of a large publisher, could expect that access to a publisher would become easier soon after crossing the 1000 unit threshold. The unknown author would now be a known quantity, which is less of a gamble for a publisher to market. At this juncture, who is now the gatekeeper of taste, of determining what is popular, or what is good, or deemed worthy of mass-distribution? The scales seem to tip a bit towards the consumer.

An equally, and possibly more important element to this equation is that a publication (or bauble, or bangle, or video, or [place your favorite consumable item here]) has a much lower threshold for success. There is a distinct reason why large publishers and content creators maintained such barriers to entry. From production to marketing, the cost of creating the content used to be enormous. But now...on demand publishing, Wikis, networking, and well-placed Public Relations can theoretically make anyone a successful small press.

How the Web levels the playing field for commerce is not limited to artists trying to sell books, movies or music. Restaurants and other entertainment venues also have their own outlets where crowdsourcing shapes the notion of what is good, and what is not. Yelp.com started in San Francisco in 2004 and rapidly spread to other metropolitan areas allowing users to form their own opinions about where to go for the best noodles or electronic dance music. On Yelp anyone could post their opinion regarding their experience at a restaurant, or coffee shop, or other cultural venue. The beauty about the crowdsourcing aspect of Yelp is that a single bad review is not going to tank the reputation of any business. There are disgruntled people who have legitimately bad experiences, but even the best venues have a bad night here and there. A pattern of bad reviews, or a pattern of great reviews is what sways opinions greater than a single poor or fantastic review. The positive and negative influences of an always present tool like Yelp has even lead to restaurant owners coming up with strategies for leveraging the effects of social websites like Yelp. (Duxbury, 2008)

This notion is one of the elements behind Robert Lai's capturing of the other $19. When the drivers of demand are no longer heavily funded advertising campaigns but grass-roots discovery of a need, crowdsourcing becomes very important. This reality is also behind the changing economics in other developing regions from India to Africa. As Jamais Cascio commented, technology, specifically mobile communication technology, is an "extraordinary boon for developing world."

In India mobile phone usage has exceeded landline usage for

telecommunications. In Africa mobile phones are not only facilitating communication, but are advancing commerce, because the providers of mobile communications in these developing regions recognize the needs the masses of their users are requesting.

Jamais illustrated that there is a...

> ...*strong correlation between the rise of mobile phone use in African countries and a corresponding rise in economic activity, because people have the ability to get around traditional middlemen. ...You have people who are living out in the fringes able to check prices for the crops that they've just picked. Do I want to take it to this village or that village? Well, the prices are better in that village. You have Ethiopian coffee growers, for the first time, able to market their coffee online – directly – as opposed to having to go through multiple layers of middlemen. In some ways we have to stop thinking of them as just phones.*

In an off-camera conversation he pointed out how the processing and storage power of the iPhone I was carrying equaled or exceeded the Apple Macintosh G3 desktop computer that debuted in 1997, which at the time was touted as a supercomputer for your desk. Not just from the standpoint of technological advancement, but related to economic advancement, in regions of Africa the demands of users motivated mobile service providers to allow users to trade minutes as currency.

Jamais continued,

> *Because the banking system has not really made itself available to poorer parts of the [African] nations, the phone companies have stepped in and allow people to essentially loan each other minutes – buy and trade minutes – as a way of trading for goods and services, as a way of giving money to someone, as a way of essentially operating a viable economy without access to tradition financial institutions.*

As access to technology increases there is no argument that these regions are going to have great influence on the direction of the global marketplace in the near future, and that they are already influencing global commerce in the creative spaces. An example if this is how much more rapidly areas of Asia are adapting and adopting new technologies.

As David Fleck said, the "community in general in Asia is adopting and embracing technology at an alarmingly fast rate."

Regarding how this will affect global economies, he continued that the growth of technology use in the Asia-Pacific Rim "is about leveraging some of the talent that's out there, that has not yet had a chance to participate in what has gone on in the European and American arenas."

David also mentioned that the many new technologies placed in the hands of many in Asia are solely fixed in a virtual environment.

> *Any time you give technology to people they're going to figure out interesting and new ways to utilize them, and benefit from them. So virtual worlds give these people an opportunity to come in and not only socialize with the rest of the world, but also create businesses that they can earn incomes, and lucrative incomes on, without ever leaving wherever it is that they reside...The reason that's important is that once again these become conduits to the global platform, the global economic platforms of the world that really narrow the gap between the rich and the poor in the sense that people who maybe were, in the real world, factory workers, or farmers, or whatever, people who were working very hard for their daily incomes, can actually come into a virtual world, earn Western currencies, and apply those back into their real life and actually increase their quality of life as a result of the economics that that brings to them. That's something that wasn't possible just a few years ago.*

Opportunists in the West may drool at the short-term prospects, but compare the potential of this prospect with the economic complacency that has settled into much of the West regarding its economic prowess, and the rise of the Japanese auto industry against Detroit will dwarf in comparison to the overwhelming effects Asia's growth will have over the US and the world economies.

Regardless of the amount of economic opportunities, there is no way that the digital universe alone will create either Utopia or cultural disintegration.

As Julian Dibbell noted, capitalism does this very well by itself.

> *Karl Marx. His analysis right off the bat was 'all that is solid melts into air,' and that was the phrase that leapt into my mind the first time I heard of a [virtual] gold farm. Here was this bizarre parody of industrial capitalism, these worker bees droning away in a factory on things that [are] completely immaterial. But it's built into the logic of capitalism from the beginning. Capitalism is a phenomenon that is constantly trying to reinvent, find the next best way to do things on a level of production – not on a social level ... not on a moral level, just the next most efficient means of doing it. So it's a constant revolution in the means of production...that was Marx's point, and that's something we see all the time. And what that means then is it's always moving towards the next best thing – not so interested in what exists now – always striving for that potential improvement and that which is pure potential is the virtual – that's the abstract. So capitalism has got to love [video] games, because games are all about 'what if we do this, or what if we do that? Let's make rules and see what happens.' It's like a marriage made in heaven – or hell – depending on your perspective.*

Section 3: Somewhere in Between

Chapter 7: Safe as Ever

Plus ça change, plus c'est la même chose. (The more things change, the more they stay the same.) – French Proverb

Barry Glassner's book ***Culture of Fear*** perfectly illustrates why alarmist perspectives of the digital universe are overblown. His book, while not focusing on the Internet, specifically, supports a larger premise regarding creative content that we all consume on a daily basis. The business of content creation, while related to what we consume, has little to do with the responsible creation of content.

One of Glassner's chapters deals specifically with the fear news outlets have driven into first time parents regarding the perceived safety of children in today's society. Reading today's headlines you might think that predators lurk around every street corner, and prowl every neighborhood in dilapidated El Caminos, waiting for that opportunity to snatch innocent children from the front lawns of suburbia.

As Glassner quotes in his book from a Hugh Downs report on ABC's 20/20, "Depraved people are reaching right into your home and touching your child." (p. 33)

Glassner, however, investigates the empirical evidence behind the anecdotal claims that our children are in peril, in one section of his book he asks us to

> *...consider a suspenseful yarn that took up much of the space of the **Los Angeles Times** entitled 'Youngsters Falling Prey to Seducers in Computer Web Crime.' It was about a fifteen-year-old whose parents found him missing...Yet when the reporter gets to the conclusion of [the boy's] saga it's something of an anticlimax. The teenager returned home and informed his parents he had not been harmed by his e-mail companion, who was only a little older than [the boy] himself. Nonetheless, the moral of [the boy's] story was, according to the **Los Angeles Times** reporter: 'Such are the frightening new frontiers of cyberspace, a place where the child thought safely tucked away in his or her own room may be in greater danger than anyone could imagine.' (p. 35)*

Glassner emphasizes, "Now there's a misleading message."

Glassner's comments stem from a note in his Introduction, "Mary Douglas, the eminent anthropologist who devoted much of her career to studying how people interpret risk, pointed out that every society has an almost infinite quantity of potential dangers from which to choose."

An anecdote Glassner gave for this, to set the tone for his book, was about the $10 million US cities spent in the 1990s to remove asbestos from public schools,

> *...even though removing asbestos from buildings posed a greater health hazard than leaving it in place. At a time when about one-third of the nation's schools were in need of extensive repairs the money might have been spent to renovate dilapidated buildings. But hazards posed by seeping asbestos are morally repugnant. (Glassner, 1999)*

Any crimes against children are morally repugnant, which is why so much time and energy is spent safeguarding these children who are already quite safe. What may be even more frightening, however, is a quote from a **Time Magazine** article Glassner provides from Republican Senator Dan Coats of Indiana, "We face a unique, disturbing and urgent circumstance, because it is children who are the computer experts in our nation's families." (p. 60)

What are parents really afraid of?

Some may argue that the hyperbole around cyber-dangers, and online pedophiles is necessary to insure a lack of apathy, and to insure that the horrors of child abduction never happen to even a single child. I agree that no family should ever have to experience the loss of a child, and every child should be able to live without such traumas, but there must be a point where we are also denuding our children of the ability to cope with danger. I'm not talking about consciously placing a child in harm's way, I'm talking about teaching children about the dangers inherent in the world, and giving them tools for coping with these dangers.

Finding anecdotal evidence about the dangers we all witnessed as children is easy to do in a safe environment. In talking to other people about this project, I often talked about how safe the Internet is when compared to our known, terrestrial existence. As a boy in second grade I was approached by someone in a car who asked if I wanted a ride home. I said "no," because that's what my mother taught me to do. She taught me to be aware, and how to keep myself safe, which helped me keep myself safe as I got older. As a teenager I was taking a shower in an open shower at a local YMCA. I had my head down, looking at the floor of the shower, when I had that feeling I was being watched. When I looked up, the only other person in the shower, a man who must have been in his late twenties or early thirties, was watching me while he masturbated. Needless to say I exited the shower at that moment, but when I brought up these stories to others, at least half of the people to whom I had spoken had stories of their own. Life comes with risks, and those risks existed long before the Internet and mobile phones.

Protecting a child to the point that eliminates all potential risks requires a bodyguard, and 24-hour-a-day surveillance; these are things that no child needs. Children also need to learn autonomy, and personal responsibility. Children, and their parents, need to learn how to live. The hyperbole, sensationalism, and fear mongering that has lead to children no longer playing in the front yards of their houses in many communities indicates to me that we have gone too far.

Glassner points out that "three out of four parents say they fear that their child will be kidnapped by a stranger," and yet incidents of abduction "according to criminal justice experts...total...200 to 300 children a year. Another 4600 of America's 64 million children [as of 1999] (.001 percent) are seized by non-family members and later returned." (p. 61)

As illustrated in one of danah boyd's earlier comments, parents restricted the radius children could travel, but children always find ways to expand their sphere of influence. The car did this in the 1950s, and the Internet is expanding the boundaries today. Following Glassner's premise regarding safety from child abduction, we have to ask about child safety on the Internet.

Recent anecdotes actually show how shifting norms are making the Internet a place of potential safety for children on the Internet. As seen by the children at Mt. Diablo High School, the shifting norms quickly assume themselves in youth culture. What is ironic, however, is that the mechanisms by which these children come about stumbling across these norms is no different than what has gone on for millennia.

As Aaron P., the Junior at Mount Diablo High School mentioned earlier said,

I have a very diverse amount of friends. People

are always trying new things that might be new to them, but for us to our ears, it's like, "World of Warcraft...you did, you did all that? It's kind of like raising the bar for people to try. Well if he did that, I've gotta bring something back to the table, you know, do something better. It's more of a masculine, gotta be the top-dog kind-of thing, so for me it's just kind of funny to watch, people try to come back with new things, and some stuff they think is better, and when you hear it you say "well you shouldn't have done that," it just drops you down lower. It's a weird cycle that people go through trying to top out things.

In a roundabout way, Aaron is explaining those adolescent games that have occurred for generations. "I dare ya..." one kid says to another. The stakes keep getting raised, until one child finally goes too far with the dare, so far that either someone gets hurt, or the collective group of peers reestablish their new norm for how far is too far, or both.

Parents either forget, or choose not to remember their own struggles at this age. Differences in culture – "real world," terrestrially-based cultures – can exacerbate the struggles between a child and his or her parents. Another one of the Mount Diablo High School students, Katie T, a Senior, described how her parents first allowed Internet access in the house, then took the access away. Unfortunately for her parents, education has accelerated its own movement towards technology so now the oldest of Katie's little sisters, who is graduating from elementary school, is learning how to research assignments on the Internet, and creating further pressure on reluctant parents to have Internet access in the home to facilitate their children's education. Fear of cyber-bullying is another of the reasons Katie's parents removed Internet access from the home.

As she described,

...that's one of the reasons we don't have Internet access: because they're afraid....I did have Internet before, but it was dial-up, and it was real slow, and I was the only one who used it, and sometimes I

*would stay on, like, for a lot of hours, and I would just
be messaging on AIM [(AOL Instant Messaging
service)], and I'd just be on these random pages – but
they'd be art pages...and they'd walk in, and it's just the
wrong moment.*

*[My parents come from] completely different
worlds. They value family time too, so they saw me stuck
to the computer for hours, and they were like 'you're not
spending enough time with family,'... and since my
sisters are now... in elementary school they're kind of
required to research on the Internet now. They're getting
introduced to using computers more. One's in fifth
grade and the other's in third. And one of my sisters --
the one that's about to graduate -- she wants Internet,
and we got into this discussion. [Katie's parents were
worried that] she might get into some of these sites, and
porn's going to pop up. First of all, Mozilla Firefox
completely blocks all those [pop ups] out. Second of all,
mainly, to get porn to pop up at you, you kind of have to
search for it.*

Katie is a good kid. She recognizes the cultural difference with
which her parents have to struggle, and chooses to work with them – as
much as a teenager can – instead of leveraging those differences to her
advantage. For many parents and children caught in the struggle of a
generation gap, technology becomes a positive device for both sides.
As Christian Renaud says,

*...it sells, because both sides want to see it. It
gives both sides an excuse from actually trying to bridge
the gap.*

He continues that…

*... it allows digital natives to look at their
otherwise concerned parents and say, "you wouldn't
understand." It's not like we haven't been saying that
for millennia, but now we have yet another way to...now
I can IM it to you.*

The difficulty for the parents in this dynamic is not the mere existence of the technology, but the paradigm of thought into which digital natives have been born. Digital natives do not expect stasis; they expect, and sometimes demand constant change and revision.

Coye Cheshire speaks to this,

> ...when we're designing a system – an online interaction space, or new online community, it's not just that the uses are emergent, but they're in flux and they are a process; there is no end-point. There is no final 'you have created the system that everyone wanted, therefore we can all just stop,' and that really scares a lot of people to think that, wait a minute, does that mean that there is no perfect version, that there is no final version of anything from my word processor to my online email system, or a particular online meeting space. That can be very frightening for some people, but I think for many of us that's not so frightening because it just points out the fact that sometimes the way we think about the technology, and the use of that technology might be somewhat limiting, that we need to be able to think of it in terms of not end-points, and starting-points, but instead think of it in terms of facilitation, behaviors, and ongoing processes of change as people use things differently, they share those uses, norms are created, norms shift, they go away, they come back. All of these different kinds of things can happen, they have to do with social interaction, the way in which we share these kinds of things; that's the ongoing human experience. If it's being played out in an online space, or an off-line space, it's still social interaction. It's still exchanging information, goods, services.

For as long as the transition continues into the wholesale acceptance of Computer Mediated Communication as a legitimate, useful, and positive mode of human interaction, we will continue to be able to find stories about the horrors of a digital existence.

December 10, 2007. Fox News.com *Online Game Meetings Sometimes End Tragically, but Phenomenon Remains Rare* (Donaldson-Evans, 2007)

June 4, 2008. CNN.com. *Naked photos, emails, get teens in trouble*. (Press, 2008)

June 20, 2008. Telegraph.co.uk. *Internet addiction is a 'clinical disorder.'* (Bloxham, 2008)

Note the qualifying phrase in the first headline, "[the] Phenomenon [of tragic circumstances stemming from online behavior] Remains Rare." Slowly, headlines and stories start to reflect the reality of life in a computer mediated space: it is regrettably similar to our "real" lives. There are good stories and bad stories.

A recent article in the *San Francisco Chronicle* notes how the fantasy lives created in places like Second Life can be therapeutic. (Parsons, 2008) I even have a friend from college who became a psychiatrist who will occasionally meet patients in his "office" in Second Life.

A recent bulletin board posting on **b3ta.com** illustrates the more normative behaviors of adults that show, much to the dismay of those afraid of the Web, that it is quite safe for children. The mission of b3ta.com is to find the "best of the web," and in the vein of content being driven by users, most of the site's material is submitted by its users and edited (monitored) by the site's operators. They have a Question Of The Week (QOTW) section where the editors asked users what they thought about kids.

Here is the transcript from an April 21, 2008 posting (humpty, 2008):

> Well, I tend to use the computer at home a bit, and every now and then people add me to their list at random.
>
> This person wasn't an exception. I ignored the "Hi M8" things and the "ASL?" requests, but no matter what, this annoyance kept on coming back. Sometimes

it'd be abusive, sometimes just annoying, but one day I was bored though, and I decided to accept the challenge of a conversation.

The grammar was non-existent, the spelling was horrific, and the phraseology was right out of Charver [sic] 101. I usually Like talking to random lasses, but this one was unreal. Stupid and mind-numbingly immature. I fired off the usual 'off the shelf' insults "work at McD's?" etc... and got a reply that I didn't expect.

"Not old enough to have a job"

oh... alright.. how old was this person? I'd assumed they were about 19...

"Fuck off.. why whould [sic] I tell you"

Because I've just been slamming the hell out of you because of your childishness, but maybe you ARE a child and I should be cutting you some slack!!

"Oh... well, I'm 11"

Riiiight. In which case I'm sorry for being mean. I had no idea you where that young

Needless to say I can't remember everything that happened or how it was said... but I'll do my best.

The conversation continued, and we spoke on and off for a few days. I had been wrong. It wasn't a girl either. It was a little lad. He told me he was in the 'web to find some friends or at least someone to talk to, and he asked where I lived. I sent a couple of google links to Swedish picture searches, and he seemed to love the idea of other countries etc. He then said that he wished he could live in another land, but he had to move to London with his mum.

"Really? Why are you moving there?"

My mum says I have to, or she'll chuck me out on the street

"Woah... that's a bit mean. She's probably joking..."

No she's not. She hits me, it really hurts.

Humpty stops and takes stock: An 11 year-old Manchestor [sic] kid has confided in me... he's unhappy, in need of mates and claims that his mum is violent towards him. 2 options... he's taking me for a ride... (look out for requests for financial support) or he's serious. No harm in talking to the lad... What could possibly go wrong?

"She hits you?"

That's not good, why does she do that?

I don't know. She said she wishes I was dead, and that I'm ruining her life.

Have you told anyone?

No, I don't want to. My sister and my mum like each other.

Right. Let's get this straight. Your mum hits you, and it makes you unhappy.. and you haven't told anyone?

Yeah...

Well that's wrong. Your mum is supposed to help you as you grow up, not hit you. You *really* need to tell someone and talk to them about it.

Yeah, but who?

Teachers. You could try telling them?

I'm not good at school, the teachers don't like me

That doesn't matter. This is FAR more important than school, and they will know that. They'll help. It doesn't even have to be one of your teachers. Pick someone you like, or one of your friend's teachers and ask if you can talk to them... Tell them everything that you've told me.

Are you sure that'll be ok?

Yes. Absolutely. That's what teachers are for. Teaching is only a bit of their job, looking after pupils is what it's ALL about.

Ok.

and I'll be here as usual... ok?

yup.

****

I heard nothing for a couple of days... then a girl named Haley added me to her list.

She started out with "You don't know me but you know my brother."

Oh shit... here we go: Kiddy-fiddling accusations..

"I just wanted to say thank you. I think you may have saved his life...."

It turned out that the day after I took the time to speak with him properly, he'd walked into school, and with a thumping heart, had walked up to his maths[sic] teacher - for whom he had some respect - and told him exactly what he'd told me.

According to his sister who'd been at home that day, police turned up at her mum's doorstep at midday and took her away. Both she and her brother were now living with their biological dad, and they were both really enjoying it.

The night that all this had happened and after social services had spoken to them both, Her little brother then went and sat on her bed and pulled his shirt up - for the first time ever his sister found out that her mum had been beating him. He was covered in bruises - all over his body. He told her about talking t me, and he told her that he's been thinking of killing himself: his classmates had surmised that this was probably his only option anyway.

So... she thanked me for being there when he needed someone, and giving good advice in a way that he could identify with it. It was a pretty cool feeling.

****

Kids might type like shit...

They may not embrace correct grammar..

They may really piss you off...

They might swear and spit...

...but underneath, some of them are just lost little kids.

Don't write them all off.. not yet."

Fear mongers might cast this anecdote off as an anomaly, but I see this speaking to where our Computer Mediated Communicating culture is going. Technology is rapidly changing, but the sociological foundation of humanity is not. We know somewhere deep down inside that we need each other, and kids know this inherently. Yes, there are going to be adults who leverage this knowledge to their own twisted ends, but this is no different today than it has been for years; it's just easier to hear about it today. What was also heart-warming about this transcript were the follow up comments to this post.

Here were the first few posts; the rest followed in kind:

-I hope this is a true QOTW [Question of the Week]

Because if it was, it's truly beautiful and has restored my somewhat lagging faith in humanity

-Crikey!

Bet you feel well proud of yourself! That's sweet, too! Deserves a click in my book!

-it is.

Absoloutly [sic] true.

Was about 18 months ago I think. Still in touch with the lad: he spent last summer living with his aunt in Spain and started to learn Spanish... Just blimmin' fantastic. =)

-Bloody Well Done

Humpty, you may be a rotten-fish mimsy vomiting scoundrel, but fucking well done.

Salutes

The replies not only show support, but also a small bit of the flavor of who are regular viewer / contributors to the site, along with a sense of the community that has grown around this small street corner on the Web.

The one that really caught my eye was "Humpty, you may be a rotten-fish mimsy vomiting scoundrel, but fucking well done."

This single-sentence reply reflects so much about not only the thread of discussion, but of the community around the site. While I may be taking this out of context, and what I'm reading may be the comments of a bunch of dilletantes (sic), the commenter at least has read enough of the "Humpty's" postings to have some opinion regarding Humpty, and despite the commenter's opinions of Humpty the commenter was civil enough to compliment Humpty on his posting and actions illustrated in the posting. In reaction to the first reply in the thread, from my own perspective, yes, this does restore my faith in humanity – that we can all get along even despite our differences – and find the time to help others in need.

Is a child really at that much of a greater risk from sexual predators on the Internet? The likelihood is not as great as many media outlets would like us to believe. At the initial onset of the Internet, where children and parents were still very unaware of what the Internet could facilitate, yes, we could assume that children were at greater risk. The differences, however, between a child's life on the Internet today versus a short five years ago, is that cultural norms quickly adapted to catch up with this new environment.

Think about email. When new adopters to email first started

using the communication medium many would type with two fingers and the cap-lock of the keyboard left on. In email norms, which had already been established by email users, this was offensive; the same as screaming all of your opinions at everyone. Established users of email did not chastise these new users, they simply informed them of the already established norms of email usage. This is an advantage families like the 13 year-old boy's from the *Frontline* story did not have. That boy and his peers were the first digital citizens in places like MySpace. The most dangerous moments – yet the one's at which the most change occurs – are on the fringes where two previously disparate worlds collide. When the 13 year-old boy from the story was involved in Computer Mediated Communication the worlds of "real-world" teenage interactions and digital interactions were just beginning to collide, or integrate, depending on your perspective. The problem with being a first adopter of any new technology is that the norms discussed by folks like Coye and danah are not yet established. But now that social networking sites are more established digital natives are already finding their third, or fourth, or fifth "cool" place. For digital natives in 2008 they're past the moment of initial intersection; they're fully immersed: the norms are established.

Talk with grade school students today and they'll tell you how crazy you have to be to chat with anyone online you don't know, and that if you do, you're looking for trouble. Grade school. Not high-school or college age kids, but children in fourth and fifth grade. I took some time to speak with children at Parkmead Keyspot, the after-school program hosted on the campus of Parkmead Elementary School in Walnut Creek, CA. In one anecdote Jasenia W. said,

> *...cuz I was, like, seven, and I went on [this online Superhero game], and this weird guy came up to me in the game; it's sort of bad some times.*

What is most striking about Jasenia's comment is the ease with which she recognized an inappropriate situation, and the corresponding ease in which she both addressed and avoided the situation. Unlike an MSNBC expose, she did not engage this "weird guy."

What we do have to remember, however, is that children at this age, while uncannily savvy about "stranger danger" awareness, even on the Web, are, after all, still kids. One child commented, "If you're online, playing online games, they can almost easily hack onto your computer, because you're connected onto the Internet, and you're on the Internet too."

The kids are right to worry about hackers – they're probably more vulnerable to hackers than to sexual predators – but the terms "hacking" and "hackers" seem to get thrown around as blanket terms for online danger. The intuition and instinct of some children is very encouraging regarding the well being of children online.

I had been trying to find out how kids think and react to strangers on the Internet, and whether any of them were in less secure environments, like chat-rooms. Here is one interchange I had with Ifoma E., a fourth grader at Parkmead Keyspot.

Ifoma said, "I make sure that I don't go on websites that say, like, do you want to have the privilege to chat on the website, because they can really be [unsafe]."

I asked, "Do you know that from the stuff that you've done, or did your parents tell you this?

She returned, "I've done that, [gone on sites with chat rooms]. You sign up and they say to give your email address and your last name, and then, like, where do you live… and then you start to chat, and you never know. Because you never know [where someone is from.] What I do now, if it's like a chat website, I just press 'block!'"

This doesn't mean that the norms that are developing through Computer Mediated Communication are the same as old norms, or are norms that will make hesitant adopters comfortable. How can changes in cultural norms be qualified? Norms are not necessarily "the average behavior," or a part of a representative bell curve as Coye Cheshire goes on to explain.

> *We see certain kinds of behaviors emerge, they're not necessarily predicted in any specific way, it's just that we see them happen, and we see people perform them, and we see other people perform them, and after a while we end up with certain kinds of behaviors that are expected, and certain things that are not expected. Now the interesting thing about norms is that you often don't know that you have a norm until you look at it when it's broken. That is, you look to see what*

happens when people do something, and people admonish them for it, for doing something, or respond...It doesn't have to be a negative kind of sanction, but it is one of those things where we notice them when people say "oh you didn't say that right. Here's how we say it here. Here's how we say this. Here's how we communicate this. Oh you just type...No, no, no. You don't type that, you type 'lol' when you're laughing at that," and all of these other kinds of things. That's how we begin to observe them. So it's an interesting thing because it happens in all kinds of communities, and groups, and cultures. So the fact that we see it happening in different groups on the Internet, and through different types of [Computer] Mediated Communication is pretty fascinating...And they continue to evolve; it's not just static, and they might differ quite wildly between different groups.

Human beings like to be noticed, and some like to be noticed more than others. We like to be heard, respected, and to feel like we matter. Communication is our easiest channel for feeling connected, and for feeling like we matter. While small-scale oral and written communication has always been accessible to everyone (from writing letters to gabbing at the local coffee shop), large-scale communication has been the purview of a select few until recently. The growth of hyperbole in our creative content could be tied to the rapid expansion in the ability of the masses to reach the masses. Or is it?

Our news media thrives on sensationalist headlines. After all, what is more compelling – more viscerally grabbing – a visual story of cute puppies, or just-in, tragedy unfolding, video of a burning car crash? The car crash wins hands down.

Dramatic writing 101 is about, well, drama. What is drama, but tension? If a scene in a book or a film falls flat it's because there's no dramatic tension. There's no room in our media culture, with outlet after outlet of media for consumption, for anything that lacks dramatic tension. Be aware the next time you are watching a TV show or movie that does not feel compelling, and observe how often a gun appears in the story – or some other threatening weapon – just when you were about to get up for popcorn, or switch the channel. Introducing a gun, or weapon is the easiest way a writer can "raise the stakes," in order to create dramatic tension.

So of course our news media thrives on tragedy. It is drama. The stakes have been raised. News, for everything else we'd like it to be, is also a business selling creative content, and each news outlet is competing to draw attention away from all of the other news outlets available on radio, TV, online and in print.

Some of the cultural reasons for the success of sensationalism come from something called the Burkian Sublime: the reason we slow down to stare at a car crash. In the latter part of the 18th century British philosopher Edmund Burke defined his notions of Beauty and the sublime in his publication *A Philosophical Inquiry into the Origin of Our Ideas of The Sublime and Beautiful With Several Other Additions*.

In discussing the sublime he said it "operates in a manner analogous to terror...that is, it is productive of the strongest emotion which the mind is capable of feeling." (Burke, 1787, p. 80)

Just as interesting as Burke's comments on the Sublime is how I found the quotation: through Google's Book Search, which contains the ongoing and growing results of its book scanning initiative, Google Print, discussed earlier.

The Burkian Sublime begins to explain how hyperbole has become the norm in the content of almost everything we consume. Anything truly striving to be unbiased is usually an academic paper, and you don't particularly see those flying off the shelves at Barnes and Noble, or listed as top sellers at Amazon.com. Sensationalism sells creative content – even news. This is why the stories of the death over a virtual sword, the beating endured as retaliation for the killing of an in-world avatar, and murder resulting from an online love triangle is so compelling and heavily carried by media outlets, while stories about a child saved from abuse are not.

We've all complained about how the evening news leads with the worst news imaginable. Where are the countless stories of people doing good in our communities? There are good people out there. Are there cultural extremes that need to be kept in check?

Of course, but the cultural extremes are not so scary when we read Coye's comments:

...it's hard to say what most people want to do, but...I like to believe that we're pretty good people. I like to think that most people are pretty good, and that behaviors we see sometimes – the negative, the stuff that isn't so great, especially in the online space, and we hear about predators, and we hear about people doing bad things, and cheating one another, and scams, and phishing, and all of the different types of behaviors that we observe – I think that...just like before the Internet on TV and the media, sometimes interesting stories rise to the top, and we tend to think of those things as perhaps more common than they actually are. I do not think that there is any evidence, just even looking at many good studies – the Pew Internet Study, and other kinds of things where...we are keeping track of what people think about the Internet, how they are using it – I don't think there's currently any really good evidence showing that people are becoming any worse, or becoming more evil, or doing anything, or behaving more in these negative ways just because of the existence of these new technologies....With the scamming kind of stuff, it's exactly the kind of thing [where]...unfortunately, we've seen that kind of behavior before. That doesn't mean that people are doing it more than they ever did before.

Coye and danah are not the only academicians dissecting all of the data that is out there. Alluding to some of Jamais' comments, and his writing in the **Metaverse Roadmap Summary**, there is a lot of data out there, and every new technological advance allows casual users and serious researchers to more efficiently and more complexly utilize the data. Two well known research power-houses – The Pew Internet & American Life Project, and the MacArthur Foundation – are among the research organizations pouring a lot of money into the study of what Computer Mediated Communication is doing to our societies (both locally and globally), into how we can leverage the benefits of these changes, and into how we can reduce the inherent liabilities. After all, every decision and advancement comes with trade-offs.

Just as technological paradigms for creating technology still mirror real world functions, behaviors in the digital universe mirror those in the terrestrial universe. Of course there are functional elements of the digital universe that facilitate the growth of what many view as anti-social behavior. Taking the position of moral police force, however, is wrought with logical flaws.

Culture is not static; it is always in flux. Linguists understand that language is a reflection of culture. Now let's look at the "language" that digital natives are adopting in blogs, texting, and IMing: a lexicon that bleeds into virtual worlds where the current primary form of communication are texting threads. F2F for "face-to-face." 2moro for "tomorrow." LMAO for "laughing my ass off." LOL for "laughing out loud." BRB for "be right back." IMHO (In My Humble Opinion.) Like anything with digital natives, the adoption of technology, and making it their own began as soon as digital natives saw a useful opportunity. This new text-based lexicon started as soon as digital natives discovered the benefits of SMS (text) messaging on cell phones.

A 2003 article in the **Times of India** (TNN, 2003) speaks to this language revolution, one so useful that it quickly made the leap from young digital natives to corporate lingo.

The article presents a hypothetical, yet plausible, corporate scenario: "When MNC exec pro Amit Das' cellphone screen flashes 'BTOYA', he knows it's time to leave everything and run towards the war room (read: boardroom). Kyunki, 'BTOYA' means 'Be There Or Your Ass'."

The article, however, also describes an issue that creates concern still today in 2008, which is dramatic since five years in technological terms is an entire era.

> *"My friends and I use SMS lingo which is strikingly uncommon. This allows us to send across SMSs without intruders coming to know what we are talking about,"* explains Neha, a student at Hindu. *(TNN, 2003)*

The issue is that lexicons are fluid in this realm, and while potentially a public means of communication – anyone could read what you are saying over your shoulder – you can send inflammatory notes about someone standing right next to you to someone with whom you have a predetermined texting code, and not worry about any repercussions, even if the person took your phone from you and looked at the text message you sent. Written language is, ironically, becoming more complex as it starts to stray from the orthodox.

A 2008 article in the *New York Times* (Holson, 2008) continues the exploration into how this issue is exacerbating the growing digital divide between parents and children. The author of the article describes an anecdotally common moment where he chastises his daughter for text messaging on her cell phone while she sits in the back of the car with her two girlfriends. Her reply, "'But, Dad, we're texting each other,' she replied with a harrumph. 'I don't want you to hear what I'm saying.'"

But is an anecdote like this the illustration of the threshold of our civilization's downfall? Absolutely not.

As the *New York Times* article continues and relates the new realm of texting – in essence Computer Mediated Communication – to the initially explosive growth in telephone use, and to then the adaptation of the car to teen culture: the same observations echoed by researchers like danah boyd. (Holson, 2008)

Every new form of technology has come with both the promise of Utopia, and the fear of cultural collapse. Elon University in Atlanta, Georgia is an integral academic partner in the Pew Internet Project. The University maintains a website called *Imagining the Internet* where they publish many of the results of their research. On this site are interesting quotations from the past where people speculated on the effects of "new" technologies. Here were some quotations regarding the telegraph (Anderson, 2005):

> *When the first transatlantic cable was built from England to the United States and President Buchanan and Queen Victoria exchanged messages in 1858, a writer for the Times of London raved:*
>
> *'Tomorrow the hearts of the civilized world will beat in a single pulse, and from that time forth forevermore the continental divisions of the earth will, in a measure, lose those conditions of time and distance which now mark their relations.'*
>
> *Authors Charles F. Briggs and Augustus*

Maverick wrote in their 1858 book 'The Story of the Telegraph':

'Of all the marvelous achievements of modern science the electric telegraph is transcendentally the greatest and most serviceable to mankind...The whole earth will be belted with the electric current, palpitating with human thoughts and emotions...How potent a power, then, is the telegraphic destined to become in the civilization of the world! This binds together by a vital cord all the nations of the earth. It is impossible that old prejudices and hostilities should longer exist, while such an instrument has been created for an exchange of thought between all the nations of the earth.'

Chapter 8: Culture and Capital

"Every group tends to set up the means of perpetuating itself beyond the finite individuals in whom it is incarnated." – Pierre Bourdieu, *Distinction, A Social Critique of the Judgment of Taste*

While learning about the fundamentals of playwriting my favorite writing professor, Betsy Carpenter, taught us about "negotiation over an object." That object can be anything from a tangible artifact like a baseball autographed by Babe Ruth, or intangible like someone's love. There's a distinct reason why the basic fundamental of "negotiation over an object" is present in every scene of any successful play (or film); in relationships, like it or not, aware of it or not, we are always negotiating over some object or another.

Those objects are a currency, which is even how writers will refer to those objects within a play. "What's the currency?" Currency is not just, well, currency as we know it; it's something that we use to barter, or trade, or exchange for things or services that we see have equal value. In haughty academic circles these non-financial currencies are often broken into three categories: academic, social, and cultural capital.

Academic Capital is a simple gauge of how far you've gone with your education. A high amount of Academic Capital would have someone not only graduate from college, but go to graduate school, and not only go to one graduate school, but two or more. Of course there are a few other elements like where you went to school that factor into the academic capital, but that "where" starts to bleed into the other currencies: Social and Cultural Capital.

Social Capital in the eyes of some cannot be purchased – you must be born into it – hence the differentiation between old money and new money, where people from old money will disparage those with recent wealth as nouveau-riche. Social Capital can, however, be earned by following the rules prescribed by the people within the social group you are trying to enter. Some people feel that only an Ivy League education will break down those social barriers when your ancestry is that of a mill-worker's son. Within Social Capital there are many other markers ranging from knowing what fork to use at a place-setting to finding one's self on the guest lists for *the* important parties in town. Some of these markers – from higher education to table manners – can also bleed into Cultural Capital.

Cultural Capital is the currency unique to the cultural group which you are in, or want to enter. It is the TV shows, and the bubble gum, the skateboarding shoes, and logo-shirts that are the markers of cool, that label someone as accepted in their group.

As I noted in Chapter 3, in the 70s there were always a few children in our grade school classes whose parents forbade them from watching television. These kids were smart. These kids were like any other kids, but that had nothing to do with their media consumption habits. What was always noticeable, however, was their lack of comprehension of certain cultural touchstones that were so important to building social bonds. The importance of having watched the latest episode of "The Brady Bunch" was the water cooler conversation of elementary school – a starting point for conversations and communication – which helped kids discover common ground. When children were not a part of that conversation, even when others wanted to include those who had not seen the show, keeping the non-TV watchers as part of the conversation was difficult.

As Coye Cheshire notes, regarding life in front of a television or a Computer Mediated Communication device,

> *Being able to share experience, to talk about, exchange our stories about the same kind of thing, that's a way in which we build all kinds of positive social outcomes, like group solidarity, and all kinds of things that are really important for a community or a society more broadly. I would never say that people who don't participate in [something], who don't watch TV, don't spend as much time on the Internet, can't be a part of those cultures or societies – ...by definition they are still a part of that society in various senses, geographically and otherwise – but the missing of shared experiences can be an alienating thing...Part of it has to do with people having a way to express something about themselves....Why would I ever want other people to know what movies I like unless I wanted them to be able to make something of that? The fact that they know that I like a lot of science fiction movies might carry some meaning for me, knowing that I'm actually betraying some aspect of my personality, even though I'm not explicitly saying 'hi, I'm a geek,' I'm still getting across that kind of information.*

The dynamic was so subtle few probably realize how much of a part of their lives that former dynamic is today. Do you watch *Survivor*, or *Lost*? Did you go to the theater to see *Resident Evil*, or *Horton Hears a Who*? How did you perceive those shows, and movies? Even among good friends using these media is a method of maintaining cultural bonds, reaffirming friendships and commonality. Video games are yet an additional medium in this mix of media consumption.

When parents withhold or attempt to restrict a child's access to video games they are manipulating the landscape for their children, but this is what parents do by either restricting access, or giving a child unlimited access to anything. Parenting, by definition, puts parents into the position of structuring the landscape in which their children grow. By limiting access to video games – whether blatantly, or subtly – parents are limiting the cultural touchstones that the child can use as social currency when finding new friends, and building upon existing friendships.

When our oldest son approached us wanting a Nintendo DS we resisted. Creating that resistance, by saying we would consider getting him one for a birthday or Christmas – delaying our need to make an immediate decision – also prolonged the time during which he did not have a DS and a few of his friends did. Six months in the life of a 1st grader is the duration of an entire era. During the time when we withheld a DS from our son, at least two "must-have" fads passed through his class. From Yu-Gi-Oh trading cards to Attacktix action figures these fads would last for about three months. During the peak of these fads not providing your child with the tools of the fad meant you were the cruel parent. When Yu-Gi-Oh cards swept through his school we were relieved since our oldest son had already had his obsession with all things Yu-Gi-Oh, he already had the cards, and finally he had other kids to play this game with.

The funny thing about the game was that it required an advanced ability to read, something none of the 1st graders we knew had yet to reach. After Yu-Gi-Oh came Attacktix: small *Star Wars* themed action figures on a fixed base, which ticked numbers off on a counter in the fixed base when rolled along a floor or table, all of which had the ability to shoot projectiles. There were inane rules, and a heightened sense of obsession, which waned about as quickly as the ebb and flow of interest in Yu-Gi-Oh. We were able to hold our oldest son at bay long enough to avoid the craze. After Attacktix were no longer the cool thing to play at school, our son forgot to call us the worst parents in the world because we never bought any for him.

So when the DS craze washed into our oldest son's elementary school, my wife and I thought his request for this latest fad was dubious. Maybe if we found ways to avoid purchasing a DS for our son the desire for one would pass in the same way as the Attacktix. How naïve. This was a handheld gaming system with amazing graphics, games kids thought were the coolest, games after which animated TV shows were produced, and that everyone at school wanted to have. What we were also forgetting was that it was potentially the ultimate baby sitter for when I needed to cook dinner, and something to allow him to feel more connected to his peers.

One day, my wife came home from Target with a small video game for the kids, at least she claimed it was for the kids. She claimed it was to delay the need to buy our oldest a DS. The small box she had brought home – a box smaller than the box for a Whopper Jr. – had a four inch high joy stick stuck in the top of the box with a bright red ball attached to the top of the stick. This small box had three chords attached that plugged into the audio and video jacks in the TV. This small box contained all of the games of the original Atari system I had always wanted, and more. The box had Ms. PacMan, which was my wife's favorite game growing up, and the first thing *she* played when she took it out of the box.

The genie was out of the bottle. My wife had inadvertently introduced our kids to gaming culture, and they loved it. They couldn't get enough of this little game. The game required batteries. We bought more as they used up set after set of batteries just on this little game plugged into the TV. And our oldest wanted more.

Just as any good little digital native asks "why not," so did our oldest son. "Why can't I play this game with my friend on his TV?" he asked. Good question. Given the technology, it's a totally illogical question, but from a child's perspective it was a good question, and the question had reference points for him. The DS was not just a stand-alone hand-held gaming system; it allowed you to "hook up."

Hooking up in the DS lexicon is using Infrared signals transferred from one DS to another that allow two DS players to play the same game at the same time. Hooking up is collaboration. Hooking up is community. Hooking up is shared experience. No longer is the young, budding gamer immersed in his or her own little hand-held world, thumbs-a-blazing, but now the young, budding gamers are immersed in their diminutive world together with others.

Ironically, collaboration is something that we have been encouraging among children over the past twenty years – a cultural mindset that has started to surface in higher education – something that college educators are wrestling with how to manage in their students.

From the previously mentioned article published in **The Chronicle of Higher Education**, (Carlson, 2005) Mr. Carlson spoke about collaborative study and the effects it has on today's college students. Without letting technology walk free from culpability, the article does imply that the focus on self-esteem infused into education over the past decades is also partly to blame for the new breed of student. The combination of the zeitgeist of high self-esteem of working in groups has lead to a group of student excited to learn, but only what, how, and when they want. Reading much of the article starts to reinforce that we are looking not at a failed education system, but at shifting norms.

Those shifting norms spread far outside of the classroom. If exposure to video games has become as ubiquitous as exposure to billboards, why not applaud a tool that takes the young gamer out of his or her own isolated world, and at least share the experience with a friend. If the words that kids are using revolved around small digital characters and their capabilities, shouldn't the children who want to be a part of these groups have some form of exposure to these hand-held digital worlds?

While exposure to many digital worlds begins in the hand-held environments, there is cross over to TV, and online media as well, which is what these children are dealing with. We as parents need to provide our children with tools to adapt to and cope with their environment, not ours.

And herein lays one of the great challenges of the day for parenting today. The battle lines do not lie simply with should you or shouldn't you allow your child to play video games. The video game landscape is a vast, sweeping region that covers everything from the small joystick-on-a-box contraptions to complex MMOs with very mature content. Video games, really, come with some very appealing wrappers for responsible parents, including soliciting the help of educational consultants such as the Learning Company *Reader Rabbit* series of computer-based video games. In the series, an animated rabbit – Reader Rabbit – guides children through different stories. In order for the children to complete the stories, the children are presented with different tasks, the levels of which are tailored to the target audience of the children: preschool, kindergarten, grade school. The *Reader Rabbit* series is designed for kids as young as four. The series helps teach everything from pattern recognition to advanced phonics, preparing kids for school, making learning fun, and acclimatizing kids to life behind a keyboard, mouse, and screen. While the tasks are well thought out, and great tools to take children away from television sets, and get children to start thinking, the games are also self-serving. You can't blame the Learning Company for creating games this way; they will not be in business long if they are not. These computer games with training wheels are fostering the next generation of game players, and possibly a new phenomena where workers don't want their work to be their play, but expect to be paid for their play.

Remember how Sony is looking ahead to interfacing their online 3D world – Sony Home – with Sony Playstations? Wiis can already interface with Internet-enabled multi-player environments. Think about Sony's increasing integration of online environments with what we've always seen as stand-alone electronics, and think about how the Wii controllers will advance in their complexity and experience, and you can only imagine how interwoven and more realistic this gaming experience will become in a few short years. Imagine playing golf with a friend who lives 2000 miles away, and having the experience be more than a reasonable facsimile. That's not here...yet...but the technological development is racing down this path.

What may be hardest for digital non-natives is to comprehend that these generations of digital natives know nothing else, something for which you can fault neither the child nor the technology producers; it is a part of what many call progress. To expand on a notion echoed by practically everyone I interviewed, the way in which children have embraced and woven technology into their lives is no different than anything we've seen before. What's different today is that the tools of communication and creativity are different, and the scale of interaction between people that these tools facilitate is larger than anything ever before conceived.

Digital media presents a very practical liberation from what once left art as the domain of the affluent. As Katie T at Mt. Diablo High School said,

> *I would be able to do [art] on an actual canvas, and an actual paint brush, but the thing is I really can't afford [that stuff]. I feel like I can let myself go more, a lot more on the computer than I can, actually handling [materials]. It's expensive too to mess up because you've got to buy a new canvas, then the paint tubes are like five bucks.*

This is the perspective of what many see as only the second generation of digital natives, where a distinct influence of the "actual" world still exists in their schooling.

With the third generation of digital natives – those in grade school right now – the influence of actual items is becoming even more woven into the realm of the virtual. As Kate K. from the elementary school said back in Chapter 3 about spending time on Club Penguin,

> *[My friend is] on [Club Penguin] and I go with her…we communicate on the phone…so we can just talk to each other while we're playing. Tell each other where we're going… You can send cards [to your friends] and you get to talk [to your friends with the cards.] You get to have different jobs. [We] play the games, and get more money, and buy stuff for your igloo, and your penguin.*

Her friend, Christine S., added "I like buying the clothes."

I asked, "If you could make [virtual] clothes [for penguins] for other people to buy, would you do that?

They both quickly answered, "yeah."

Christine emphasized, "that'd be awesome."

My nephew may be on the upper end of the spectrum of Club Penguin users, but he is not an outlier on the bell curve of Club Penguin penguins. He has played the games in order to earn enough in-world money to buy tons of stuff for his penguin and igloo, and buy Puffles mentioned earlier. My nephew is not a pack rat. All of the things he has purchased in Club Penguin he has purchased with a purpose. At each change of season, or at the approach of significant holidays, he redecorates his penguin and igloo accordingly. He has holiday decorations and accompanying Santa-esque clothes for the penguin during Christmas. He keeps shamrock designs and leprechaun garb for the springtime near St. Patrick's Day. This boy has put a lot of thought into this online life that he maintains for this little penguin. And like many gamers, Club Penguin is not the only online world in which he takes so much care in maintaining his avatar.

While Coye Cheshire and Julian Dibbell note that culture is accelerating, that acceleration also infringes upon what some may view as the accelerating maturation of children.

Coye says, "It's quite plausible that the change that we have seen in the technology – what it facilitates – is a level of interaction that arguably hasn't been seen before, or at the same scale."

But, he continues,

> *There are similar processes. I don't think that the Internet... [has] created brand new things that we have never seen before [behaviorally], but the construction of how those things operate, and the scale in which they happen, and the implications of the speed in which they're occurring, the size of the systems that we're talking about, that does have very large implications, and I think we have to acknowledge the importance of that. Some very large-scale interactions. It's not that people couldn't, for example, contribute to public good before – we've been talking about collective action for centuries, really, at least the fundamental problem of collective action is something that goes farther back than even centuries – the idea that lots of different people are contributing to something, but then each of us having incentives that may not lead us to contribute to that because we'd rather have everyone else do the work. But now we have the facilitation of massive amounts of communication on the Internet that allows many different groups to participate, many different*

people to discuss, and contribute, to share in different types of systems, and that really is interesting. Accelerated is one way to put [what is happening.] But it's also just the sheer size of these systems leads to the production of certain kinds of things that arguable we couldn't have done before at the same speed.

Look at kids and you will see what Coye is talking about. Children are not growing up any faster than before; in elementary school they are as immature as we were when we were young. What's different is the amount of content they consume; it's massive. And even if parents are vigilant in filtering that content, with such large volumes of content bombarding children, a certain percentage will get through that is more mature than a parent wants to have directed at children. Even if parents work valiantly at limiting a child's exposure to mature content (violence, or more mature language) children still encounter other outlets that allow mature content to seep into subconscious. And then there is the conduit that young children have used for millennia: other kids and siblings.

What concerns many parents about online access is the difficulty in policing a child's access to the Internet: a seemingly limitless portal to unfiltered content. The one element of an online existence I have yet to address here is sex: not sex from a predatory perspective but the dearth of sexual content anyone can find on the Web. One reason for avoiding sex until this point is the attention that this titillating topic drags away from the other more insidious and beneficial elements of the Metaverse, and Computer Mediated Communication in general

Sex, however, has to be addressed. For many people sex is the elephant in the living room that isn't even addressed in our terrestrial existence. Think of the fights that occur regarding sex education in public schools. But as we know, sex sells, and not just from the advertising agencies on Madison Avenue. In the most hard to grasp scenarios for non-digital natives, sex on the Web is no longer pictures of extreme fetishes, or galleries of video clips of hard-core pornography. As we saw in the story of the love triangle, the Web can facilitate sexual fantasy for two or more consenting adults through simple text chats, or singles looking to find fun, or long-term monogamous relationships through the myriad of dating sites targeted to all of the varying degrees of relationship intentions. From free postings on Craigslist.com to playing the age-old role of matchmaker on eHarmony.com people can find real connections with real people that lead to real sex. They can also find this real sex with real people through their avatars.

Though Linden Labs, the makers of Second Life, do not actively promote this aspect of their virtual world, some of the most populated islands within their virtual world are the ones focused on sex.

As Julian Dibbell elaborates,

> *...people have sex via their online characters here, and it actually produces physical reactions in their bodies, and the sensation on some level of connecting with another person.*

This all dovetails perfectly with the research by Byron Reeves and others. If we experience the same endorphin rushes from the simple visual reactions we have with another person's avatar, why wouldn't that also go for what is a very visual experience for men: sex?

Of course this is going to blow the mind of parents who are already afraid enough of letting their pre-teen and teenage kids loose on the Internet. It makes earning a driver's license appear like a minor threshold in a child's development. The sex, while easy to target and brand as one of the seven deadly sins run amok on the Internet, and one of the principal reasons to loathe this new technology, plays a surprisingly minor role in the virtual worlds, and online games of significance.

Julian continues even further,

> *When I went back a second time...looking at the real game-type spaces I was surprised to see that there was no sex...None of these night owls were really spending a lot of time getting it on with each other. But then I realized we don't need the sex any more...we have the money. And if I can say to an outsider, well, yeah, you don't care about this virtual sword, but this virtual sword will fetch $500 on eBay, then people start to sort of go 'oh, OK.'*

So what is the real danger to children? Which content is the most troublesome? That question is better left to parents and child psychologist. The current reality, however, is that children in first grade are gaining access to online virtual worlds already explicitly designed for teens and older. How are they explicitly designed? The only way to participate in Internet MMOs is to register a username. During the registration process practically every MMO available asks for the age of the participant. Where a potential demographic crossover can occur, where young adolescents and children may have interest in the games, the age of the participant is a required field. When children are honest, like mine thankfully are at this moment in their lives, once the user enters a birth date dating the user as under 13, the game restricts access to the site. One MMO, GoPets, goes so far as to track the IP address – the physical location on the Internet on which the computer is seated – and block future, potentially devious attempts to register a user from that IP address. After all, kids are savvy. A second grader will quickly figure out to what birth date to enter in order to register for an MMO where the minimum age to play is 13.

Kids can also easily find out usernames and passwords their parents use.
I asked a group of fourth and fifth graders "are there ways to figure out your parents' usernames and passwords?"

Non-chalantly they answered "yeah."

One girl, Jessica G., noted, "it's in a book. My parents have a book, and I found it."

But parents should not be worried about this. Communication is paramount with children, and when parents deliver consistent messages, children listen.

One boy, Robert V., noted, "You shouldn't go on bad websites. If my parents caught me then my friends wouldn't see my face for like a year. She wouldn't let me go to school."

Ifoma E., mentioned earlier, continued with the explanation. "She wouldn't let you go outside. She would just probably go like 'this is what you do for now on: you go to your room, you eat dinner in your room, and you go to sleep.'"

From a company's perspective, having the thirteen year old threshold for MMO users is more than a legal language hoop for companies seeking to insure that underage users are not being exposed to inappropriate language. The majority of companies consciously pursuing users under thirteen also implement other more rigorous safeguards to insure the safety of the children using the sites, and the mental well being of the parents. Neo Pets, an MMO like Webkinz run by Nickelodeon, has six full-time staff members dedicated to monitoring user activity. The staff is on the lookout for everything from cyber-bullies to sexual predators. Some companies take these efforts to levels that would make even the most hesitant parent comfortable.

As of July of 2007 Habbo Hotel, a Swedish firm running a pre-teen and teen-centric virtual world, which has users in 19 different countries, has 300 of its 500 employees dedicated to monitoring the behavior of its netizens. Part of their strategy is to have staff in all of the countries where they have user presence. The reason for this is the company understands how language and culture relate. Language used in one country may have a totally different understanding and context in another country. Having local site monitors insures that users are neither unfairly accused of wrongdoing, nor allowed to slyly pass-off incidents for simple misunderstandings. And most of these sites are serious about their enforcement of their policies.

My oldest son was playing on Club Penguin when he threw a snowball at another penguin that he did not know. Just like in the real world, throwing snowballs at friends is good fun as long as everyone is having a good time and wants to participate. If you were walking past a group of children you did not know having a snowball fight, and one of the children threw a snowball at you, you might have your feelings hurt by the action. Well this little penguin had its feelings hurt from being hit by my son's snowball, and the penguin reported my son's penguin. My son was barred from Club Penguin for 24 hours.

He was devastated. He thought at first that he'd never be let on Club Penguin again. He wished he could find the little penguin he had hit and apologize. He really learned a lesson about consequences – all from throwing a virtual snowball.

Many savvy yet concerned parents, those already understanding of the communication power of the Computer Mediated Communication, will place the family computer in a very public space in the home. Watching how quickly norms shift, yet how quickly children adapt to these norms, the question arises regarding the effectiveness of the public, family computer. Digital natives are going to adapt to, utilize, and modify technology in ways that parents may never expect. Hiding the path of websites visited is very easy to do once you learn how. Flipping from one window in a computer – one that may contain content a parent does not want a child to view – to a window containing parentally safe content is, pardon the cliché, child's play for digital natives raised on **Reader Rabbit** and DS games.

The best parental defense against inappropriate use of technology by children is the same one that parents have always had: communication. But here is where the difficult element of parenting reveals itself. The cultural contexts for children and parents are so far afield that parents must work terribly hard at making sure the communication they use with their children is effective. To use my previous example about interpretations of words, we parents need to know what cup our children are seeing in their mind's eye when we use the word.

Starting our conversations with our children early is the best preventative measure we can take. Listening to the words of the fourth and fifth graders from Parkmead Keyspot in response to a question I had regarding how involved their parents are in their use of technology:

I asked "who has a computer in their room" and half of them raised their hands.

Robert V. added,

> *The thing is, whenever I'm trying to sign up for something, my dad always peers over my shoulder and he's like 'what are you signing up for?' And I'm like, 'oh, just something that is a game.' And he's like, 'did you check it before you did it?' And I said, 'No.' And he checks it, and he says, 'they'll try to hack your computer if you do this,' so I can't do it.*

I asked another question, "Would you go on your computer and [try to go on a site that your parents said not to go on], because there's pretty easy ways to make sure they don't know that you're on it?"

They all nodded 'no."

One girl, Arianna M., answered, "I'd never do that."

Ifoma E. added, "especially like if you have a computer in the living room, your parents would just walk by, because a lot of parents they're all over the house, and sometimes they'll stop to see what you're doing, so they'll walk past you and see something on the screen that's inappropriate. Like on IMVU [an online 3D chat site] there's a website..."

I interrupted Ifoma asking "you go on IMVU?" The reason I asked was because the site is an online chat site targeted to older teenagers. One look at the site's home page would be enough to give the parents of a fourth-grade girl a heart attack.

She answered, "I nearly did, and it says like, 'you can get a boyfriend and stuff,' and I really didn't like that because it's actually not for little kids."

The uses of technology, and the games our kids play are no longer going to be relegated to hand-held games that we know are safely not linked to anything else, or home-based computers in which we can forensically determine what they've been doing if we are motivated enough. Like counter-intelligence efforts, a parent's attempts at circumventing and preventing a child from participating in unsafe activities will never amount to the effectiveness of preventative education.

Every child needs an experience where they get kicked off of Club Penguin for something relatively innocent. Once the dynamic between parent and child becomes one of cat and mouse then the child will always have the upper hand. Such is the advantage a digital native has over his or her parents because as I noted in Chapter 7, Republican Senator Dan Coats of Indiana said, "We face a unique, disturbing and urgent circumstance, because it is children who are the computer experts in our nation's families." (Glassner, 1999) Such is the curse, however, of the digital native, but it is not something of which parents need to be afraid.

What is frightening, is that just like any relationship based on communication, there is a moment where the parents have to trust, and to allow the child to develop a sense of self, and autonomy, and allow the child some privacy. This is really what any child is seeking when s/he is texting friends from the backseat of a car.

danah boyd details takes a unique look at privacy:

> *Privacy is a really interesting thing, because privacy in technology is a zero or one byte: on, off.*

Privacy in everyday life is not experienced that way...at all. Basically what constitutes privacy is the feeling that you have control over the context, control over the audience, control over the information, and you understand how far it will flow, who you are telling it too, etc, and trust is a really reinforcing factor. Space is a huge element of privacy, because we can understand that speech cannot go beyond the walls. The way the architecture of the Internet works all of our social, easy understanding around privacy doesn't work. So it's really interesting to see the three ways that people try to actually create privacy in a meaningful way online. The first is by trying to build technology barriers: walls, basically virtual walls. So, for example, trying to make themselves unsearchable by putting up fake information...Inevitably, for various corporate and personal reasons this collapses at every level. Just because a wall gets built...replicability, searchability collapses in all different ways. ...The next way to cope with privacy is to demand social boundaries...My favorite example of this is, a teenager says to [her] mother 'but it's my space,' and [the] mother says, 'but it's public,' and teenager retorts 'but it's my space,' and if you hear this refrain it sounds a lot like 'it's my bedroom,' 'it's my house,' 'it's my bedroom,' right? Which is the same argument over many, many generations. So this demanding of social boundaries is really functional at a sort of interpersonal level, [and] never works in American society when you have power dynamics, because adults inevitably think they have control over teenagers, bosses think that if they can get access, they should have rights to it, everybody thinks that just because you can, you should. The next sort of coping mechanism, which is the primary adult coping mechanism, is 'ostrich.' If I don't see you, you don't exist. I'm going to ignore you. I don't see you. I'm not coping. It's such a painful way [to cope.] Adults are undeniably naive with this; they just don't cope. The problem is they read teenagers performances on these sites, and they read into them all sorts of horrible things, because they're projecting their own views of what those images are. This has been a classic contention between adults and children around sexuality...going fifty years now, strong, which is that 'oh my god, you can't wear that out of the house, what are you signaling?' Because that signal is read differently for different generations. Teenagers will talk a sexual talk long before they've actually experienced a lot of what they're talking about. They're working it out, and there's a lot that we do for working things out through talking about it.

What is amazing is that so much of privacy revolves around issues of communication. Possibly because we understand that if someone can see what we're doing then we've relinquished that certain sense of visual privacy, but having someone inadvertently see you seems far less invasive than having someone overhear a conversation you did not mean to have as public. Think of kids and texting, and the ability to code text messages so that even if they are viewed by people the messages are offending, there is a chance that the message might not be understood. Something as advanced as the modern-day smart phone is facilitating the advancement and growth in complexity of a millennia-old form of communication: written words. To rebel against this notion you could argue that the language that teenagers are using in text messages is nothing like formal written English.

Discomfort with the changes in lexicon are understandable, but the culture of texting has really nothing to do formal written English. In most texting, and IMing situations users chuckle when they see writing that is overly formal and formally grammatically correct. It's really no different than coming to terms with the concept that we no longer use Sanskrit in day-to-day conversations.

Soon that hand-held device we know as the cell phone or smart phone will do far more than SMS text messages, and IMing, or do simple Web surfing, or take pictures, or capture video, all of which is already available today. There will be full-blown virtual worlds in which any user will be able to traverse, visit with friends, and make new acquaintances, while also targeting and tracking the terrestrial, geographic location of each of those other users. Access to basic forms of Augmented Reality will soon seep into these mobile devices.

Think of anything that can be related to something else through technology and you can imagine there is a technologist out there somewhere figuring out how to make those connections happen. Augmented Reality, while initially introduced through games and entertainment venues, will have profound effects on the terrestrial world in which we operate. That disturbing intrusion is already happening today.

Jamais Cascio recounts,

What we are getting really close to is the capacity to build the kinds of technologies that would let us – and again this is going back to the Augmented Reality types of technologies – to see the world with 3D appearing virtual images walking around amongst the [already existing] physical world – existing side-by-side with the physical world, and actually we see people doing this right now. There's something out there on the Web called 'AR Quake.' Quake was an old 3D shooter video game… What people have done is take that source code, build it into an Augmented Reality system with a map of their university, so that they could run around with their little controllers, shooting at monsters that jump out of windows, or, pop out from behind the physical walls, that, of course, only the people with those goggles on can see those monsters, which, of course, leads to really zany scenes of people running around shooting at things that aren't there, at least as far as the 'normal' people are concerned. But for the people living in that space, that augmented space, those monsters seem plausibly real. And as these technologies improve, the ability to create images that seem plausibly real gets better and better. Now the question then becomes, 'who has control? Who has control over the ability to insert plausibly real images into our augmented reality?' Is it something that only happens if you load a program locally? Then you have pretty good control over it. If you want to do be able to [stream and

*provide access to an Augmented Reality] through the
Web, then it starts to get interesting because people will
then figure out ways to insert things into our field of
vision that we don't necessarily realize are 'fake.'*

Despite the amazing prospects of Augmented Reality, the
Metaverse, and Computer Mediate Communication as a whole, what
we cannot forget is that at this stage in their development – whether
they are terrestrially tangible, or virtual – these things with which we
engage are mere tools that can only do what we ask them to do.

Christian Renaud made a flippant yet accurate statement,
"Nobody ever expected anybody to be beaten to death with a shovel."

We cannot demonize tools that are used in ways that their
designers did not intend. These tools of Computer Mediated
Communication that come in all forms – from cell phones on which
teenagers are texting, to immersive 3D virtual worlds where people are
trying to convert virtual work and virtual currencies into a real world
living – are nothing more than facilitators.

Coye Cheshire notes,

*Now we have the facilitation of massive amounts
of communication on the Internet that allows many
different groups to participate…the sheer size of these
systems leads to the production of certain things that
arguably we could not have done before at the same
speed….We need to think of this in terms
of…facilitation, behaviors, and the on-going processes
of change. As people use things differently, they share
those uses, norms are created, norms shift, they go
away, they come back. All of these kinds of things can
happen. They have to do with social interactions, the
way in which we share these kinds of things. That's the
on-going human experience. If it's being played out in
an online space or an off-line space, it's still human
interaction. It's still exchanging goods, services.*

This flux in an online world is really nothing new, it's just that the Computer Mediated Communication tools facilitate an acceleration of age-old human interactions. This explanation does not eliminate the question of how parents will manage a child's exposure to these growing and ever changing technological tools. This explanation does not even deal with how adults protect themselves. What about those already sentient adults getting themselves into trouble like the Marine in Chapter 1? Some adults will see these technologies as threats, while others will see them as opportunities, and based on these perspectives parents will either enable or restrict the access children have to new and established technologies. The degree of access a parent allows is neither good, nor bad; it is a parenting choice, just as parents in the 70s made parenting decisions when limiting or restricting access to television, and parents in seventeenth century Puritan communities restricting access to dancing.

But what of the parents whose children actively seek access to, or who are allowed free access to video games? Are these parents not providing enough guidance, or boundaries for their children? Are the freedoms provided to these children going to lead to the proliferation of the kind of cultural parodies seen in the remake of *Willie Wonka and the Chocolate Factory*? In the film, one of the children to win the golden ticket was so obsessed with video games that while he was hyper-focused, and hyper-competitive, he also lacked any notion of civility.

This begs the question of what is civility? The definition of civility not only wavers from culture to culture (say between America and Western Europe), but within sub-cultures (think Midwest, versus East Coast). Cultural difference is at the heart of much of the fear of gaming culture, digital natives, and the foundation of the generation gap. Within the digital native gaming sub-culture the definition of civility is changing. The changes, however, should not shock anyone since the definition of civility is always changing. Simple cultural visual cues are always changing. Think of hat-wearing among men. Not only was hat wearing practically requisite among many cultures, there were protocol to follow. No one would ever imagine wearing a hat indoors fifty years ago, but go to any elementary school today and you will find young boys wearing baseball caps in all manners, and wearing their caps inside the classroom. They wear their baseball caps indoors, and in front of a teacher while she is conducting a class.

But are the Internet, or hand-held, or home console video games to blame for this? These media are not the shifts in norms; they are new tools to facilitate this behavior. Ultimately we – the people who participate in our society – are responsible for determining norms.

The exciting thing for gaming culture and Computer Mediated Communication in general, is that the rules are still being created as the environments are still very new in human terms. In technological terms we are already seeing at least the third generation of 3D immersive online games (some could argue more), and countless iterations of console games. The online, MMO environment, however, is the most intriguing as it is the environment where people can commune. Some may see this communing as metaphorical, but for those who participate, it is very real

The advent of the wheel probably did not come with the same cultural baggage since its propulsion for the first few millennia was based on already known quantities: people, or horse, or oxen and other ruminants. It's easy, however to imagine the fear and promise that came with advents like sail power. What lay downstream? What lay over the horizon? Textual illustrations and imagery of the fear and promise of the travel made possible by water craft continued in popular literature throughout the eighteenth century, and many would argue through today. From sea serpents, to the Bermuda Triangle, to today's digital horizons some fear the unknown, and others are drawn towards it.

Chapter 9: Creating the Future

'Someday we will build up a world telephone system, making necessary to all peoples the use of a common language or common understanding of languages, which will join all the people of the earth into one brotherhood. There will be heard throughout the earth a great voice coming out of the ether which will proclaim, "Peace on earth, good will towards men."'

– AT&T chief engineer and Electrical Review writer John J. Carty [from] his "Prophets Column" in 1891," quoted from Imaging the Internet website maintained by Elon University / Pew Internet Project. (Anderson, 2005)

The beginning of 2008 saw the stock market reacting to economic issues that some blamed on technology. Financial services companies had come to rely on economic models run on computers that forecast trends, risk, and potential profits. Paraphrasing from an earlier Christian Renaud comment, technology currently only does what we tell it to. We have yet to reach the point of The Singularity mentioned in Chapter 3, where human thought and consciousness has inalterably changed because of technological innovation. True technological innovation is still the realm of PhD students and intellectual prodigies.

This man-behind-the-curtain scenario makes it very easy for corporate heads in a financial crisis to hide behind the black box of technology when the issues that brought the US economy to its financial crisis had nothing to do with technology, but with human hubris.

From an Alvin Toffler perspective, has technology surpassed human kind's ability to process information? Perhaps there is more access to information today than ever, but how many people fifty years ago actually consumed every article in the Sunday newspaper? Whether focusing on the hunt, or working in a cubicle, human beings have always had to deal with filtering the information coming in from their environment. The rapid progress in information dissemination through new technologies has just meant that humans are now adapting to this new environmental element, developing new skills to either process more information at less depth, or to filter out more information in order to concentrate on one single topic.

But what about the kids in the back of the car texting to each other? What about the child texting, IMing, playing an MMO, and doing homework all while listening to his iPod? What about the college student boldly texting on his cell phone while his professor lectures? Yes, these digital natives have new skill sets that most older adults will never attain, or don't want to possess, but what does this mean for our culture? Doesn't it point to the unraveling of society as we know it?

Absolutely.

Society is always unraveling. That's what cultures do. Invent. Reinvent. Adapt. And with these changes come the changes in language, and not only the changes in language but the changes in how and where communication occurs.

The Journey of Man on ***PBS*** proposed that the languages spoken by indigenous tribes in Africa are descendents of those original languages spoken by our hominid ancestors, but a language like English is light years from such ancestry, which has nothing to do with technology but with cultural adaptation. Is one better or worse than another? Absolutely not; they are both perfectly adapted to the cultural needs of their owners. The human animal is extremely adept at adapting to the ultimate intangible tool: language.

danah boyd points out,

> *Talk to people in China, and they have all of these games, talking at this level, and knowing that they're talking at [a different] level. You've been watching it within Burma, talking about literature when we're really talking about the political state of things. American society [currently] just reads at one level. If it's at the surface level, that's what it is. Although teenagers are actually starting to find all of these ways of saying things at multiple levels, which says that [learning how to speak on different levels is] powerful...and that people can learn. The interesting thing is actually how gay men have been using social sites, and dating sites in particular...if you're in a gay dating site you can say whatever you want. You can out yourself. Duh, you've outed yourself by being there. But the interesting thing is how people will create multiple levels within [the site]. So, for example, you could say that you were submissive or dominant as a sort of identity marker within a profile. Instead people start using capital letters and lower case letters to tell that story at multiple levels, because there's so much invested in gay culture in reading between the lines they don't want to give it up. So you now have a generation that's growing up, and one of their coping mechanisms ... is learning to write between the lines, learning to write at multiple levels, basically in code. It's code that's far more mature than the pig-Latin of childhood. It's more, saying something at a surface level that's saying one thing, that's a direct jab at that friend at another level. So it doesn't necessarily read that to the*

teacher who might see it, but it reads that to the other people. These multiple layers are a beautiful way of [coping]. And it's a far more nuanced way than most adults coping mechanism.

So the teenagers of today are learning new skills in communication, skills seemingly lost in mainstream American culture over the past centuries. Language and its use is not going away because Computer Mediated Communication tools – whether text-based IMing tools or visually-based virtual worlds – are, at their foundation, tools for communication. Language is growing in importance. And because the Web has all of its specifically targeted corners where a person can be their true, authentic self, we're heading for a Utopia of self-expression. Right?

As danah boyd stressed,

...that's what we have to push against – Utopic ideas. There [is this] magnified sense of self that's frankly not that different from putting on make-up – finding the coolest clothes and going out to a club. You know, the make-up makes me look a little younger, a little cooler,...high heels are constructed to make me look younger, there's all of these moves that are about a certain kind of magnification of reality, that are not necessarily Utopic, I can be anybody I want to be. What you find is that people don't actually want to be anybody that they want to be, they just want to be cool, by whatever markers of cool amongst their social groups matters. And that's one of the reasons why you have to distinguish between mass use of the technology, and in many ways marginalized, and outcast's use of the technology. And if you look at new technologies at every turn, the first adopters are always the self-defined geeks, freaks, and queers – the marginalized of society who desperately want to find others like them, and are seeking to have this sort of freeing of their physical constraints, but mass use doesn't look like that. If we look at social network sites, Friendster early adopters

were geeks (bloggers), queers (mostly gay men living in New York), and freaks, self-defined freaks (primarily at the time it was Burning Man attendees living in the San Francisco Bay Area). These were populations who... were marked as outcasts but also saw themselves in many ways as part of this sub-culture that were happy to be on the margins of society [(quote, unquote)]. But they moved themselves in a way where they found others like them. As you watch social network sites grow, what you find is that people reproduce the social networks of their everyday life. A MySpace network of teenagers? That's them with their friends from school. Sure they might be trying to [be] friends with the porn queens, but by in large there's a virtual appreciation for the distance,...they're accountable off-line. That picture they put up: they're accountable the next day they're at school, they're accountable the next day at work, [and] they're accountable when they go to the club on a Friday night. It can't be that much of a discontinuous representation.

Well we're at least agreeing that human beings are good at adaptation, and The Singularity is a possible outcome somewhere out in the future. Our continual technological developments, therefore, must mean that we are driving ourselves towards an eventual Utopia where Computer Mediated Communication facilitates the complete integration of human interaction and on-demand data, where it changes us to such a degree that the norms of tomorrow are indistinguishable from anything we could ever imagine today, and the norms create a perfectly egalitarian world of free markets and pure meritocracy. Right?

Not so fast.

Expanding on what Jamais noted in Chapter 3, Utopia in terms of a Singularity, of the adaptation of the human species careening down this path to human / machine hybrids is unlikely because…

...we tend to focus, when talking about a Singularity, on the gadgetry, on the robots, and AI

[(Artificial Intelligence)], and implants, and all the kinds of devices that might make something like that possible. And what we really need to be looking at, and where I think the really interesting stories get told, are around how are people changing? Are relationships changing, whether it's technologically enabled or otherwise? How are our capacities to communicate and understand each other changing? What are the implications of these seemingly simple devices? Something that records everything? Well that's very straight forward, we have similar kinds of things at work already. But what happens when that's in everyone's hands? Well, there are some really significant changes to help people interact with each other. And all of these devices, all of these technologies, all of these evolutionary developments – this co-evolution of society and technology – the unanswered question remains, the unanswered questions remain to be, 'How do we live with this!?' And not in a 'we have to reject this sense,' and not in a "we're eager to have it sense,' but in a 'how is this changing our relationships to each other?

Along with adaptation comes decisions and choices.

News media outlets, however, will keep the drumbeat rolling on emphasizing the difference between generations, further elevating the currency of both sides. The new currencies we see evolving in Computer Mediated Communication are neither frightening, nor Utopic. We've seen this before. The new virtual currencies – from text messages to virtual gold – are no different than the claw game from the penny arcades referenced in the Introduction. Those games were a real currency played out through a game. The child who was best at the game could either horde his winnings, or trade them for other goods and services. And whether or not s/he traded any of the winnings, that child had the social and cultural capital of being great at the claw game. Simple metaphor, but it's no different than what's playing out on cell phones and PC screens today.

News reports and popular opinion regarding Computer Mediated Communication will ebb and flow for some time to come.

Many people involved in the industry, who see nothing but the promise, have these new tools elevated on a pedestal of promise while they espouse the degree to which Computer Mediated Communication will continue to reduce the distance that separates people and disparate cultures. Many Industry professionals are blind to anything except for the promise of the future. Thankfully there are publications like the Virtual Worlds News that have reporters like Joey Siler who vigilantly report on how virtual worlds are developing, including headlines like *71% of Facebook Users Have Never Heard of Second Life* from August 6, 2007. (Siler, 2007) This illustrates that even in a digital universe a cultural separation exists between digital natives, and even first adopters. Just because a digital native uses one website or another does not mean that s/he is adept at accessing all of the crannies of the digital universe.

There are many unexpected norms that have arisen in this new space. One funny one came from a woman I interviewed at the Virtual Worlds Conference in San Jose in 2007. The woman, Theodora Sites, was a researcher for Nicholas Research in New York, NY, a market research firm specializing in virtual worlds.

The thing she noted that most surprised her was that

> *...people want to eat in virtual worlds. People want to go be around food, and go to restaurants, and have dates. They need something to do. That kind of activity, and that kind of ritual needs to exist, and so Kraft went into Second Life and they had a big success because people actually did want to eat the food.*

She furthered her view to the personal level.

> *My own personal really shocking experience was that I kept going to American Apparel [in Second Life] and then I went to American Apparel in real life. I kept going in Second Life because they had tacos, and that was appealing...because you could hold the tacos, and you could smoke cigars, and it was something you could actually do. And then I went to the American Apparel in my town and was suddenly like 'whoa' because I felt like I knew it, and it was familiar, but it was familiar from*

the virtual world. So being able to immerse yourself that much into a virtual space, and then kind of be scared by it in real life. When I see people [in real life] and they look their avatars, and they look like someone I know as an avatar. That cross-over that you didn't think would happen – I don't spend that much time in the virtual worlds – but to have that kind of cross-over... or even Scions. I saw them virtually before I saw them in real life, so I thought that they were a virtual car that existed in Second Life, and I didn't understand that they existed in real life, and then I saw one on the street and I [jumped.] It's not something [about which] you're conscious that it's happening, and once you see it....it's a strange phenomenon.

Gerry Bell, a partner at the firm where Theodora worked, expanded on her comments about food.

You don't normally think about food in terms of it as a social connector, but when you're able to step back and analyze the purpose of food in real life versus virtual life, you can see that they're the same. Because people socialize around food, for a lot of people food is love, and so the attraction to food and the attraction to food eating situations is sort of a manifestation of the normality that people want to bring with them to [places like] Second Life. It's just like people getting jobs in Second Life, and people want to have a routine in Second Life. In a manner of speaking they want to replicate their lives, but in another way they want to make their lives different. So it's all about what works for different people and their interaction with the medium.

You could also argue from Gerry's comment, it's about how people interact with each other, not just the medium. This mirrors Christian Renaud's observation that technology today does nothing more than mirror our terrestrial needs and actions. Think of the egalitarian nature of the Internet that proponents love to highlight. The vox populi has always fought for egalitarian stature. This can be seen in everything from the French Revolution to simple modern publications like Zagat's. Of course the metaphor is a bit of stretch, particularly on the ideological level when mixed with politics. After all, there are so many more variables involved in politics, like simple human behavior. Many simple illustrations do appear in modern times, however, like Zagat's. The Zagat's website describes that the original Zagat's handbook was a consumer survey-based publication for finding places to eat, drink, stay and play. The publications today have "ratings and reviews based on the opinions of over 300,000 surveyors from around the globe." (Zagat, 2008) With this terrestrial example as perspective, the development of sites like Yelp was just a matter of time.

People will herald the Web and Computer Mediated Communication as the gateway to a new, prosperous, and totally democratic future, while other people will demonize them. Regardless of perspective, the tools of Computer Mediated Communication involve people. This is about human interaction, and until human beings begin to radically change patterns of behavior that sociologists have recognized repeating themselves for millennia, we're in no danger of Computer Mediated Communication either withering away from lack of relevance, or taking over our terrestrial existence.

*   *   *   *

A recent multimedia gallery on boston.com highlighted excerpts from Emory University English Professor Mark Bauerlein's book *The Dumbest Generation, How the Digital Age Stupefies Young Americans and Jeopardizes Our Future*. The title says it all, and on initial inspection Mr. Bauerlein's arguments are understandable. His arguments are also a stark illustration of how the fear of the unknown, or incomprehensible, is alive and well, fueling everything from the generation gap to xenophobia. Boston.com highlighted eight of his arguments (The Boston Globe, 2008):

1. They make excellent "Jaywalking" targets

2. They don't read books – and don't want to, either.

3. They can't spell

4. They get ridiculed for original thought, good writing

5. Grand Theft Auto IV, etc.

6. They don't store the information

7. Because their teachers don't tell them so

8. Because they're young

What I found most interesting about the promotion of this publication on Boston.com was the sensationalist nature of it; it fits in perfectly with Christian Renaud's and danah boyd's comments on the different generations wanting to maintain a generation gap, and with Coye Cheshire's affirmation that what sells creative content is hyperbole. How many people would pick up *IMHO*, or watch the companion documentary film if I started with the sentence, "there's really nothing new here."

The French proverb really does hold true when it comes to technology and culture: the more things change, the more they stay the same. Through all of the technological innovations, people are still born, people still die, and we all care about what it is to be human.

What's convenient about the list of eight reasons is their distillation of many of the fears that exist of this generation of digital natives. This is not a criticism of Professor Bauerlein's book, but a commentary on the means by which the book's publisher chose to promote it.

Let's start with reason number one: Jaywalking fodder. Jay Leno started the Jaywalking bit on the *Tonight Show* when he took over the helm of the show in 1992. The bit had Jay taking a camera and microphone onto the streets of Los Angeles and asking people questions about current events or history, questions that anyone with a middle school education should be able to answer. Jay started this bit four years before the word Internet became ubiquitous in American culture, three years before Amazon.com, nine years before Wikipedia, four years before ICQ (the first popular, mainstream Instant Messaging service), and one year before AOL 1.0 for Windows was released. Technology did not drive the current path to ignorance our society is on. Jay Leno recognized in 1992 that American ignorance was funny, and that no one was ashamed of it; it was just funny.

Scour media outlets today and you might find that Jay's bit might also speak to other cultural nuances. A show currently airing on the Discovery Channel called *Cash Cab* has the host of the show driving around the streets of New York in a taxicab. When he picks up his fares he asks if they want to participate in a game show where they need to answer questions on history and current events. If they get three wrong, they get tossed out of the cab before reaching their destination. If they reach their destination without getting three questions wrong they get to keep the money they earn for each correct answer, and some of the questions are not questions you hear every day. Is it that folks in New York City are that much more engaged in retaining and exchanging knowledge, or is it that one comedic bit wants to showcase the idiocy of ignorance, while the other wants to showcase and reward the retention and use of knowledge? Oh, and most of the participants on *Cash Cab* appear to be under 30.

There's also something to the American disdain for the educational elite that has existed for centuries. Often portrayed as stuffy, and out of touch with the common man, American pathos has ridiculed and been disparaging towards book-learning for years, even going back to the days of Daniel Boone. Nerds. Geeks. Absent minded professors. Bumper stickers reading "my kid beat up your honors student." American ignorance was alive and well long before Computer Mediated Communications came into play.

Reason number two: they don't read books – and they don't want to, either. To criticize this element of a sub-group is short sighted, and not taking into account the changing ways of distributing information. Though not completely accurate, it would be akin to someone at the start of the twentieth century saying "they don't ride horses any more, and they have no interest." Is there something special and different about sitting down and spending time with a good book printed on paper that you can touch and smell as well as see? Of course, but to create a list of cultural touchstones that have faded in the past century in the wake of "progress" would take a lengthy book of its own. True, reading habits are changing, but people are still consuming vast amounts of text. We no longer read books printed on a Gutenberg press. Does the delivery vehicle matter? Look at the growing success of Amazon's Kindle and Sony's eReader devices for reading electronic books. We can at least agree that the how is changing.

Reason number three: they can't spell. This is like a schoolyard spat breaking down to someone saying, "oh yeah, you smell." Do we in America still spell tire "tyre?" Spelling changes. Language changes. It always has, and always will. Seeing that linguists note language is a reflection of culture, culture is always changing, and so is language.

Reason number four: they get ridiculed for original thought, good writing. Similar to the Jaywalking reason, if a researcher was seeking examples of sub-cultures in a place like MySpace (which is where the author found his examples of writing ridicule), s/he would be just as fortunate finding examples of users who ridicule others for poor writing. But we also have to remember that language and communication is culturally specific. As we age, our cultural references become more narrow as there is not as much time to leap from group to group, from school to friends, from one group of friends to another, from teachers to parents, all of whom have different expectations of what is normal for communication. And MySpace is supposed to be a place for friends to hang out since parents have restricted the distance that their children can travel to go hang out. The language, by definition, is going to be totally counter to what is expected of them in school, or by their parents. It is a place for them to immerse themselves in their own culture.

Reason number five: Grand Theft Auto IV. As we've seen, Computer Mediated Communication facilitates, it doesn't create. Grand Theft Auto is not popular because of technology. This younger generation is not dumb for wanting to play this game, and for making it the most popular gaming title in history. The popularity of the game is a reflection of shifting norms.

Reason number six: They don't store information.

Christian Renaud pointed out, "Einstein said, 'never commit to memory anything that you can write down.'"

Just as the distribution of information is changing, so is the storage of that information. Until ten years ago, physically storing the amount of information available in a place like the Internet today would seem ludicrous. Now, not only is there more information available through the Internet than in the entire Library of Congress, but it is all searchable and indexable, and the ability to search and index is becoming easier and easier, and more and more intuitive for users. So the question is no longer "why are they not storing information," but "why should they store information?"

As Christian continued,

> Why do I have to memorize all of this stuff if it's always available to me? Give me the actual logical reason why I need to memorize a Shakespeare sonnet. If there's some cognitive benefit from that, then you can explain that to me, but actually rote memorization...look, I have the entire corpus of human knowledge available to me on-demand, why would I need [to memorize it]?

Reason number seven: Because their teachers don't tell them so. Again, this argument started long before Computer Mediated Communications, and is solely a product of the changing academic paradigms. I'm OK. You're OK. Oh, it's not if little Johnny got everything correct, it's if he feels good about himself. Driving in the car yesterday I was carrying my two sons, and two of my oldest son's friends. One said "I'm not going to do swim team because all you get is ribbons. You don't get any trophies."

I asked, "so you don't play baseball because you love the sport, you play it because you get trophies?"

I asked this of the one boy in the car who I know loves baseball, and even he answered, "well...yeah!"

Of course I had to stop myself with the "when I was a boy..." rant, but I couldn't stop myself totally. I did say to them, "gosh, if you could you guys would want to get a trophy for going poop in the bathroom!"

They were between six and eight years-old. They thought this was hilarious. I was going for the effect. I was going for the laugh. I was also dead serious.

If we criticize this younger generation for these outlandish expectations, we have no one to blame but ourselves. Parents made this mess, not the kids.

Reason number eight: Because they're young.

There's the perfect synopsis for why there's a generation gap. Culture is changing, and fear comes from those unwilling to adapt. How many people know how to rub together sticks to make fire? How many people know Morse code? Skills appear. Skills fade into memory.

Christian Renaud noted

...as a species we use tools, and we create tools. We create tools we don't have any use for...Underlying it all, though...there's this deep human need to connect...Society's gonna change. Technology is going to be digested at different rates of speed in different parts of the world. People are going to use it for things that the creators never meant it to be used for. These are standards. This has always been the case. Nobody ever expected anybody to be beaten to death with a shovel, I'm pretty sure. [There are] unintended consequences, and that's going to be true of all these things. The most recent bogey man was that terrorists are using virtual worlds to get together and plan things, and therefore the NSA needs to wiretap them. And I'm pretty sure they said the same thing about telephones, and telegraphs. There was an old H.L. Menken quote...'the government's job is to keep ... the populace alarmed

with imaginary hobgoblins,' and that's going to be true of society in general. The generation that can't grasp what's going on as far as change is going to say, 'nope, I long for the idyllic (probably fictional idyllic) days past,' and the current generation is going to go, 'I don't know what your problem is. What's wrong with the Charleston, or jazz, or rock, or...virtual worlds, or texting?

Reality will never be usurped by computers...not totally. Has Computer Mediated Communication taken over, and radically transformed how we connect to each other? Yes. Absolutely yes, and there are still miles to go before the limits of the transformation are reached, but this is no reason for fear. Lack of fear, however, is also no reason to lack skepticism. What is normal behavior in these new communications media is still to be determined.

As Jamais Cascio says,

When we think about technology as a social construct, when I talk about this relationship between technology and society, it's important to recognize that it's not simply technology as a driver for social change. Society determines what happens with technology. Technology is a cultural artifact. And what that means is that it isn't just some amorphous institution that develops 'Technology,' with a capital 'T.' In fact it's every day decisions in markets, every day decisions in universities, every day decisions by individuals, and families, and communities about what they want, what they're willing to accept, what they want to reject, what kinds of life they want to lead, and so, we can't just let technology be imposed upon us. That is an abdication of our responsibility, as citizens, and as members of a civilization. We have this choice, it is incumbent upon us to make a choice about what technologies we want to adopt. It is incumbent upon us to make a choice about what technologies we want to develop. Whether we are researchers or lay-people, we will say 'this is where I want to go, this is where I don't want to go, and here's why.' And that requires education, that requires a bit of foresight, and a capacity to imagine, because ultimately the only way we're going to survive as a planet, as a civilization, this century of extraordinary transformation – environmental, technological, social, political, economic – the only way we're going to survive is to look ahead. We can't get trapped in the present. We have to be looking forward to see what the possibilities are, not to make predictions. Not to say 'this is what will happen,' but to say 'this is what might happen, how do I want to avoid that? This is another thing that might happen, how can we steer in that direction?' We have to recognize that, in the words of Bruce Sterling, 'the future is not a destination, it's a process.' It's in all of our hands to contribute to that process.

*   *   *   *

Let us remember that when we react fearfully it is most likely because we are reacting to something that we don't know. Conversely when we are overly excited about something there are things about it that we are choosing to not see. If you remember back to the Introduction you will remember that this project started over a question my wife and I had regarding buying a Nintendo DS for our oldest son, we were afraid of what effect the game would have on him. We were ill-informed, and ignorant to what hand-held games represented for young, elementary school kids.

As you might imagine, we acquiesced and eventually purchased him a DS. We soon purchased one for his little brother, and as a family we finally took the major step and purchased a Wii. The games play a well balanced role in the household: baby sitters, learning tools, currency with which my wife and I can coerce our children into doing things. And there are more times that I have to dust all of the games from lack of use than peel our kids off of the games from over-use. Most importantly, we are trying to be a part of the process that is creating our kids' futures, and remember that technology is not responsible for the actions of our children; we are.

Works Cited

A&M Records, Inc. v. Napster, Inc., 239 F.3d 1004 (9th Circuit 2001).

Abramson, M. (2008, January 6). Cyber schmooze is not just for geeks. *The Sunday Times* , p. A34.

Anderson, J. Q. (2005, August). *1830s-1860s-Telegraph*. Retrieved March 5, 2008, from Imagining the Internet: A History and Forecast: http://www.elon.edu/e-web/predictions/150/1830.xhtml

Appleyard, B. (2007, April 22). *The web is dead; long live the web.* Retrieved April 23, 2007, from Times Online: http://technology.timesonline.co.uk/tol/news/tech_and_web/the_web/article1673425.ece

Bailenson, J. (2008, April 27). Learning from the Virtual You. (A. Seabrook, Interviewer)

BBC. (2005, March 31). *'Game theft' led to fatal attack*. Retrieved March 2, 2008, from BBC News: http://news.bbc.co.uk/2/hi/technology/4397159.stm

Bloxham, A. (2008, June 20). *Internet addiction is a 'clinical disorder'.* Retrieved June 20, 2008, from Telegraph.co.uk: http://www.telegraph.co.uk/news/uknews/2152972/Internet-addiction-is-a-%27clinical-disorder%27.html

Buckleitner, W. (2008, May 8). When Web Time Is Playtime. *The New York Times: Circuits* , p. C8.

Burke, E. (1787). *A Philosophical Inquiry into the Origin of Our Ideas of The Sublime and Beautiful With Several Other Additions.* London: J. Dodsley.

Buskirk, E. V. (2007, Novembet 5). *Listening Post*. Retrieved May 30, 2008, from Wired.com: http://blog.wired.com/music/2007/11/comscore-2-out-.html

Carlson, S. (2005, October 7). *The Net Generation Goes to College.* Retrieved March 20, 2008, from The Chronicle of Higher Education: http://chronicle.com/free/v52/i07/07a03401.htm

CRD. (2007, November 8). *The launch forum of 'China Virtual economy District'.* Retrieved May 5, 2008, from CRD.gov.cn: http://www.crd.gov.cn/en/index.asp

Dell, K. (2008, May 12). *How Second Life Affects Real Life.* Retrieved May 12, 2008, from Time in Partnership with CNN: http://www.time.com/time/health/article/0,8599,1739601,00.html?cnn= yes

Dictionary, A. H. (2006, January 1). *Avatar -- Definitions from Dictionary.com.* Retrieved August 22, 2008, from Dictionary.com: http://dictionary.reference.com/browse/avatar

Donaldson-Evans, C. (2007, December 10). *Online Game Meetings Sometimes End Tragically, but Phenomenon Remains Rare.* Retrieved December 10, 2007, from Fox News: http://www.foxnews.com/story/0,2933,316333,00.html

Duxbury, S. (2008, June 27). *Restaurants learn to Yelp.* Retrieved July 28, 2008, from San Francisco Business Times: http://sanfrancisco.bizjournals.com/sanfrancisco/stories/2008/06/30/sto ry1.html

Fowler, G. A., & Qin, J. (2007, March 30). *QQ: China's New Coin of the Realm?* Retrieved July 22, 2008, from The Wall Street Journal: http://online.wsj.com/public/article/SB117519670114653518-FR_svDHxRtxkvNmGwwpouq_hl2g_20080329.html?mod=rss_free

Fram, A., & Tompson, T. (2007, November 13). Parents say no to computer games. *The Contra Costa Times (Associated Press)* , p. C2.

Glassner, B. (1999). *The Culture of Fear.* New York, NY: Basic Books.

Heffernan, V. (2008, April 27). *Sepia No More.* Retrieved April 30,

2008, from NY Times Magazine:
http://www.nytimes.com/2008/04/27/magazine/27wwln-medium-t.html?pagewanted=1

Holson, L. M. (2008, March 9). Text Generation Gap: U R 2 Old (JK).
New York Times , p. D1.

Horrigan, J. B. (2008, July). *Home Broadband Adoption 2008.*
Retrieved August 10, 2008, from Pew Internet & American Life
Project: http://www.pewinternet.org/pdfs/PIP_Broadband_2008.pdf

Howe, J. (2007, June 5). *Crowdsourcing: tracking the rise of the
amateur.* Retrieved May 24, 2008, from Wired Blog Network:
http://crowdsourcing.typepad.com/cs/2007/06/index.html

humpty. (2008, April 21). *Question of the Week.* Retrieved June 10,
2008, from b3ta.com: http://www.b3ta.com/questions/kids/post145106

Ivan, T. (2008, September 25). *NPD: Games Increasingly Popular
With Girls.* Retrieved October 15, 2008, from Edge: http://www.edge-online.com/news/npd-games-increasingly-popular-with-girls

Johnson, C. Y. (2007, June 11). New social website tempts the
inquisitive. *The Boston Globe: Business & Innovation* , p. E1.

Kugel, S. (2007, October 9). A House That's Just Unreal. *The New York
Times: House & Home section* , pp. D1,8.

Lee, J. (2008, April 14). *WOW hits 1 million concurrent users in
China.* Retrieved May 20, 2008, from gamesindustry.biz:
http://www.gamesindustry.biz/articles/wow-hits-1-million-concurrent-users-in-china

Levander, M. (2001, June 4). *Where Does Fantasy End?* Retrieved
April 26, 2008, from Time.com:
http://www.time.com/time/interactive/entertainment/gangs_np.html

NCTA. (2008, January). *Statistics.* Retrieved May 20, 2008, from
National Cable & Telecommunications Association:
http://www.ncta.com/Statistic/Statistic/Statistics.aspx

Parsons, C. (2008, July 13). *Second Life offers healing, therapeutic options for users.* Retrieved July 13, 2008, from SF Gate: http://www.sfgate.com/cgi-bin/article.cgi?f=/c/a/2008/07/13/LVL211GP5C.DTL

PBS. (2008, January 22). *Growing Up Online.* Retrieved February 20, 2008, from PBS.org: http://www.pbs.org/wgbh/pages/frontline/kidsonline/

Pisani, J. (2008, July 22). *IPhone Will Be A Game Changer For Games Industry.* Retrieved July 23, 2008, from CNBC: http://www.cnbc.com//id/25797789

Press, A. (2008, June 4). *Naked photos, emails, get teens in trouble.* Retrieved June 4, 2008, from CNN: http://www.cnn.com/2008/CRIME/06/04/naked.teens.ap/index.html

Quint, B. (2004, October 6). *Google Print Expands Access to Books with Digitization Offer to All Publishers.* Retrieved May 10, 2008, from Information Today: http://newsbreaks.infotoday.com/nbreader.asp?ArticleID=16357

Renaud, C. (2008, July 18). *Functional 'Augmented Reality'?* Retrieved July 24, 2008, from Christian Renaud's Weblog: http://www.christianrenaud.com/weblog/2008/07/functional-augmented-reality.html

Reuters. (2004, May 12). *Survey: Video gamers getting older, heading online.* Retrieved May 20, 2008, from USA Today: http://www.usatoday.com/tech/news/2004-05-12-gamer-demographics_x.htm

Schiesel, S. (2006, May 2). *Entropia Universe Players Can Cash Their Online Earnings at the A.T.M.* Retrieved May 5, 2008, from The New York Times: http://www.nytimes.com/2006/05/02/arts/02entr.html

Schifferes, S. (2006, August 3). *How the internet transformed business.* Retrieved May 15, 2008, from BBC News Online: http://news.bbc.co.uk/2/hi/business/5235332.stm

Schonfeld, E. (2008, January 10). *The Wall Street Journal Edges Towards Free*. Retrieved October 15, 2008, from TechCrunch: http://www.techcrunch.com/2008/01/10/the-wall-street-journal-edges-towards-free/

Shachtman, N. (2008, May). The Shield. *Wired Magazine* , pp. 142-148.

Siler, J. (2007, August 6). *71% of Facebook Users Have Never Heard of Second Life*. Retrieved July 30, 2008, from Virtual Worlds News: http://www.virtualworldsnews.com/2007/08/23-of-facebook-.html

Slotnik, D. E. (2007, July 23). Cute Friends to Collect, and Plug In to the Internet. *The New York Times: Technology* , p. C7.

The Boston Globe. (2008, May 14). *8 reasons why this is the dumbest generation*. Retrieved May 16, 2008, from boston.com: http://www.boston.com/lifestyle/gallery/dumbestgeneration/

THOCP. (2007, July 11). *UNIVAC*. Retrieved August 25, 1008, from The History of Computing Project: http://www.thocp.net/hardware/univac.htm

Thompson, C. (2007, January 22). *22-year-old killed after being drawn into deceptive Internet relationship, authorities say.* Retrieved May 29, 2008, from Associated Press: http://www.ap.org

TNN, I. (2003, August 12). *F2f with SMS lingo*. Retrieved May 2, 2008, from Times of India: http://timesofindia.indiatimes.com/articleshow/126807.cms

Tofler, A. (1970). *Future Shock.* New York: Bantam.

USA. (2007, March). *Families and Living Arrangements*. Retrieved March 15, 2008, from US Census Bureau: http://www.census.gov/population/www/socdemo/hh-fam.html

WebMD. (2006, July 3). *Detox For Video Game Addiction?* Retrieved May 15, 2008, from CNS News: http://www.cbsnews.com/stories/2006/07/03/health/webmd/main17739

56.shtml

Wikipedia. (2008, July 20). *Atari 2600*. Retrieved May 20, 2008, from Wikipedia: http://en.wikipedia.org/wiki/Atari_2600

Wikipedia. (2008, August 21). *Avatar*. Retrieved August 22, 2008, from Wikipedia: http://en.wikipedia.org/wiki/Avatar

Yadegaran, J. (2008, January 8). Virtual affairs of the heart. *The Contra Costa Times* , p. D1.

Young, S. (2008, April 14). When teens go wild, we blame the Internet. *The Contra Costa Times* , p. D1.

Zagat. (2008, January 01). *About Us*. Retrieved July 30, 2008, from Zagat: http://www.zagat.com/about/

Acknowledgements

As with many creative projects, this one could never have happened without the love and support of many, many people. First and foremost, my wife, who, through all of her own vocational struggles, still always found time and energy to pull me off of the couch and out into the world. Without you, this book and its soon to be released film would never have even begun. I love you, and thank you.

Though the next thank you is cliché, it is honest. To my parents, Roderick and Janice Lavallee; you have always believed in me, even in the dark days when I did not believe in myself. Without you I would not be the man I am today.

To Stephen Sheffield, my creative partner for more than a decade now, I give my thanks not only for this book's cover, but for the years of encouragement and belief in my ability.

To Atticus Fisher, my dear friend and writing partner: you are the one who always knows how to remind me to find the story, to keep focused, and to keep it regular. Now we just have to get you off your butt to publish your own magic.

To Simon Miller, my old friend, thank you for following me on yet another journey of discovery. I don't know what I'd ever do without you.

To Becky Rodia Schoenfeld, who found me on Facebook ten years after leaving graduate school, you have been a wonderful advocate for my work, and for self publishing. Thank you.

A special thanks to Lee Graham for his publicity expertise, and for guiding with a firm and compassionate hand down the path of successful self-promotion

For all of the material that you have read in this book there are many people I have quoted and a few who I never quoted, but from all of them I learned much about virtual worlds, computer mediated communication, and about people.

Special thanks go to Christian Renaud and Jamais Cascio. The two of you have been wonderful resources, confidants, and after this much time I'd think I would be able to call you friends.

To David Fleck, and Damon Hernandez, special thanks for the help you gave me at the outset of this project, pointing me in the direction of people who have a real pulse on what is going on with computer mediated communication.

To Coye Cheshire, thank you for stopping by my house on the way home after teaching, and for being such a welcoming soul, and for truly setting me straight on the true meaning Computer Mediated Communication.

To danah boyd, thank you for meeting me in such a unorthodox interview space in Milwaukee, WI, of all places, and for being such a lovely voice of clarity.

To the rest of you quoted in this book – Gerry Bell, Julian Dibbell, Theodora Sites, Reuben Steiger, and Sibley Verbeck – thank you for your time, for your patience, for the wealth of knowledge you all possess, and for being such wonderful people. Some of you welcomed me into your homes. All of you took precious time out of your hectic schedules. Thank you.

To those of you not directly quoted – Henrik Bennetsen and Damon Hernandez – thank you for all of your insight into a world that is moving faster than most can imagine.

To the kids of Parkmead Elementary School's Keyspot Program – Andrew T., Ben A., Jessica G., Jenna G., Lauren M., Cameron C., Solomon A., Arianna M., Robert V., Ifoma E., Jasenia W., Christine S. and Kate K. – thank you for being great kids, and helping me have a fun time seeing where our future is headed. We're going to be just fine.

To Geoff Fontanilla, thank you for all of your effort coordinating the Keyspot interviews. To Basic Books, a warm thank you for giving me permission to use excerpts from Barry Glassner's *The Culture of Fear*. To Janna Anderson, Jeff Howe, Elliot Van Buskirk, and the staff of b3ta.com, a special thank you for the permissions to use your works.

And finally to the rest of my friends and family who were patient enough to read my blog postings, early drafts, or to just hear me ramble about the project – Michelle Lavallee, David Beiser, Matthew Bauer, Bruce Meakem, Matt Magne, Mark Jacobstein, Steve Haber, Mike Isip, Paul Santucci, Dick Weyand, Paul Morris, Danny Benson, Josh Eckhaus, Laura Halpin, Chinnavuth DeMonteiro, Sophie Lay, Steve Welty, Craig McLeod, and Monica Mele to name a few – thank you. Knowing that you surround me makes me see that I am truly blessed.

www.ingramcontent.com/pod-product-compliance
Lightning Source LLC
Chambersburg PA
CBHW031404180326
41458CB00043B/6610/J